Springer Series on
ATOMIC, OPTICAL, AND PLASMA PHYSICS 65

Springer Series on
ATOMIC, OPTICAL, AND PLASMA PHYSICS

The Springer Series on Atomic, Optical, and Plasma Physics covers in a comprehensive manner theory and experiment in the entire field of atoms and molecules and their interaction with electromagnetic radiation. Books in the series provide a rich source of new ideas and techniques with wide applications in fields such as chemistry, materials science, astrophysics, surface science, plasma technology, advanced optics, aeronomy, and engineering. Laser physics is a particular connecting theme that has provided much of the continuing impetus for new developments in the field. The purpose of the series is to cover the gap between standard undergraduate textbooks and the research literature with emphasis on the fundamental ideas, methods, techniques, and results in the field.

Please view available titles in *Springer Series on Atomic, Optical, and Plasma Physics* on series homepage http://www.springer.com/series/411

Fernando Haas

Quantum Plasmas

An Hydrodynamic Approach

 Springer

Fernando Haas
Universidade Federal do Paraná
Curitiba
Brazil
ferhaas@fisica.ufpr.br

ISSN 1615-5653
ISBN 978-1-4419-8200-1   e-ISBN 978-1-4419-8201-8
DOI 10.1007/978-1-4419-8201-8
Springer New York Dordrecht Heidelberg London

Library of Congress Control Number: 2011934487

Springer is part of Springer Science+Business Media (www.springer.com)

To Rejane and my parents

Preface

The monograph is intended to provide an overview of the basic concepts and methods in the emerging area of quantum plasmas. In the near future, quantum effects in plasmas tend to be unavoidable, specially in high density scenarios such as in the next-generation intense laser-solid density plasma experiment or in compact astrophysics objects. Moreover, quantum plasmas are in the forefront of many intriguing questions around the transition from microscopic to macroscopic modeling of charged particle systems in general. In addition, the methods used for quantum plasmas can be readily translated to related areas which are currently pushing forward the frontiers of plasma science. This is valid, in particular, when using Wigner function tools for strongly coupled ultra-cold and Rydberg plasmas.

In recent years, the *quantum hydrodynamic model* became popular as a simplified but not simplistic approach for quantum plasmas. In particular, the nonlinear aspects of quantum plasmas are much more accessible using a fluid description, in comparison with kinetic theory. The aim of this book is to give an account of the basic developments on the hydrodynamic paradigm for quantum plasma problems, readable by a broad audience. Therefore, the proofs and mathematical calculations are given with some detail, usually not shown in the papers of the literature, due to brevity needs. Hence, some "tricks" needed to achieve most mathematical results are discussed here and there. This is the case, for instance, in the derivation in Chap. 2 of the evolution equation for the reduced one-particle Wigner function. Further examples, as well as new developments, appear in the exercises at the end of each chapter.

In the same context, in the Introduction, a very brief account on classical and quantum plasmas is offered. Here, the differences and similarities of the classical and quantum cases are stressed. We hope that in this way the book can become valuable for readers not necessarily fully acquainted with theoretical plasma physics and quantum mechanics. However, some level of knowledge is presumed: basic statistical mechanics and nonrelativistic quantum mechanics. Some familiarity with plasma methods is also advisable, although not mandatory.

The monograph is not intended to be encyclopedic. Rather, the chosen topics reflect the particular experience of the author. Nevertheless, there is a scientifically arguable reason for the sequence of contents, so as to make the book as self-contained as possible. Hence, the first chapter is an overview of classical and quantum plasmas. Chapter 2 is dedicated to the basic kinetic model for quantum plasmas, namely the Wigner–Poisson system. Here, the essentials on Wigner functions and electrostatic quantum plasmas are discussed. Chapter 3 dealt with the first attempt to a fluid model for quantum plasmas, based on the quantum Dawson (or multistream) model. The nontrivial peculiarities of the stability problem of streaming equilibria in quantum plasmas are analyzed. In Chap. 4, the quantum hydrodynamic model for plasmas is derived. The merits and intrinsic approximations of this approach are addressed. Chapter 5 is dedicated to the quantum ion-acoustic waves as described by the quantum hydrodynamic model. Chapter 6 generalize the quantum hydrodynamic model to include magnetic fields. The associated magnetohydrodynamic equations are then derived. Chapters 7 and 8 apply the quantum hydrodynamic equations to the nonlinear interaction between Langmuir and ion-acoustic waves in a quantum plasma. The corresponding quantum Zakharov system is considered in one (Chap. 7) and three (Chap. 8) spatial dimensions. In Chap. 9, a moment method approach provides an alternative macroscopic description for quantum plasmas, in the electrostatic and electromagnetic cases. The above sequence of topics goes in the sense of increasing complexity.

Along the history of plasma physics, most nature and laboratory plasmas fit in density and temperature regimes so that classical descriptions can be safely employed. With the ongoing miniaturization and the experimental assessment of new parameter regimes, however, the need to take into account quantum effects in many-body charged particle systems is becoming a reality. Hopefully, this monograph can be useful against the prejudice according to which plasma science is necessarily classical. In this manner, we expect to encourage researchers to work in this basically unexplored emerging field, whose consequences are for the moment largely unknown.

I want to express my gratitude (in alphabetic order) to Serge Bouquet (Paris), Antoine-Claude Bret (Ciudad Real), Gert Brodin (Umeå), Bengt Eliasson (Bochum), Leonardo Garcia (Porto Alegre), João Goedert (Porto Alegre), Paul-Antoine Hervieux (Strasbourg), Giovanni Manfredi (Strasbourg), Mattias Marklund (Umeå), Waleed Moslem (Port Said), Refaat Sabry (Mansoura), Padma Kant Shukla (Bochum) and Jens Zamanian (Umeå) for the collaboration and support over the years, without which this book would not be possible. However, of course they are not responsible for the mistakes in it.

Curitiba, Brazil   Fernando Haas

Contents

Acronyms

V_s	Adiabatic speed of sound
V_A	Alfvén velocity
κ_B	Boltzmann's constant
μ	Chemical potential
Γ_C	Classical energy coupling parameter
λ_B	de Broglie wavelength
λ_D	Debye length
$\mathbf{E}$	Electric field
$-e$	Electron charge
n_0	Equilibrium particle number density
E_F	Fermi energy
λ_F	Fermi length
T_F	Fermi temperature
v_F	Fermi velocity
k_F	Fermi wavenumber
c_s	Ion-acoustic velocity
Ω_i	Ion cyclotron velocity
σ	Longitudinal electrical conductivity
$\mathbf{B}$	Magnetic field
$\hbar$	Planck's constant h divided by 2π
ω_p	Plasma frequency
Γ_Q	Quantum energy coupling parameter
c_s	Quantum ion-acoustic velocity
ϕ	Scalar potential
c	Speed of light in vacuum
T	Thermodynamic temperature
q_t	Test charge
v_T	Thermal velocity
λ_F	Thomas-Fermi length
ε_0	Vacuum permittivity
$\mathbf{A}$	Vector potential

Chapter 1
Introduction

Abstract We compare the essential physical parameters of classical and quantum plasmas. The screening properties in degenerate and nondegenerate plasmas are discussed, in terms of Thomas–Fermi and Debye lengths, respectively. The coupling parameters associated with particle correlations are considered for classical and quantum plasmas. For classical plasmas, the average kinetic energy per particle is of the order of the thermal energy, while for dense systems it is of the order of the Fermi energy. A general approach toward fluid models deduced from kinetic descriptions of charged particle systems is proposed. This chapter is finished with brief historical notes on quantum plasma physics.

1.1 Classical and Quantum Plasmas

A plasma is generally understood as a many-body system composed by a large number of charged particles whose behavior is dominated by collective effects mediated by the electromagnetic force. Here, the word "collective" designate phenomena determined by the whole ensemble of particles in the system. For instance, wave motion in plasmas has a collective character. The self-consistent, electromagnetic mean field in a plasma is also a result from the collective properties of the system, and so on. In an opposite manner, the behavior of neutral gases is much more influenced by short-range interactions, or collisions.

The collective aspects of plasma physics are due to the long-range of the electromagnetic force. In this regard, binary collisions where the velocity vectors of each particle dramatically changes orientation in a small spatial volume and in a short time are unlikely in a plasma. Instead, the cumulative effect resulting from many small scattering angle collisions gives the more significant trajectory deviations in plasmas.

One of the basic plasma parameters is the temperature. According to Big Bang theory, in the origin of everything the temperature was so high that no atoms or

F. Haas, *Quantum Plasmas: An Hydrodynamic Approach*, Springer Series on Atomic, Optical, and Plasma Physics 65, DOI 10.1007/978-1-4419-8201-8_1,
© Springer Science+Business Media, LLC 2011

molecules could have existed. Hence, the corresponding fully ionized gas was in the plasma state, so that 100% of the Universe was a plasma, the so-called quark-gluon plasma. Later, as the temperature cooled down, the matter was able to assume its other well-known states. Namely, the gaseous, liquid, and solid states appeared. However, today most matter is still in the plasma state. Sometime people assigns a definite fraction like 99% of the Universe to be made of plasma, but such estimates are obviously hard to verify.

The ability of the particles of a system to associate with themselves increase in the same measure as the temperature becomes smaller. In this context, with the cooling of a fully ionized gas, some fraction of positive and negative charges can combine to form atoms. In this case, we would have a partially ionized plasma. As the temperature decrease, eventually the degree of ionization is so negligible that the system can be treated as a neutral gas, composed by weakly interacting atoms and molecules. There is no sudden transformation from the plasma state to the neutral gas state. Unlike the phase transitions from gas to liquid and from liquid to the solid state, what happens is more in the form of a gradual change of the degree of ionization. The gas–liquid and liquid–solid transformations are also produced by a cooling of the temperature. In particular, in the phase transition from liquid to solid the temperature is so small that a periodic lattice can appear, with the ions occupying a more or less fixed position in space.

To associate a meaning to the expression "small temperature," it is necessary to compare the thermal and interaction energies of the system. In this way, a coupling parameter can be defined, as the ratio between thermal and binding energies. Indeed, in the transformation from plasma to neutral gas, and then to liquid, and then to solid, the particles of the system became more and more attached specially due to the organizing character of the Coulomb force. Hence, in our definition of a plasma, we can add the requirement of a sufficiently high thermal energy, in comparison to the binding energy. However, this choice of nomenclature exclude the so-called strongly coupled plasmas, where collective effects are also decisive.

Actually, strongly coupled plasmas have more in common with a liquid than with a standard weakly coupled plasma [12]. As examples of strongly coupled plasmas, we have the interior of giant planets like Jupiter, solid-laser ablation plasmas, plasmas in high pressure arc discharges where the thermal and ionization energies are similar, and cold dusty plasmas. On the other hand, the systems found in space plasma physics, astrophysics, controlled nuclear fusion, and ionospheric physics are all weakly coupled.

Besides temperature, the density (number of particles per unit volume) is a fundamental parameter of a plasma. In a first guess, we would imagine that extremely dense plasmas should be necessarily strongly coupled. This would be the case, for instance, for the plasma in massive compact astrophysical objects like white dwarfs, where the density is as high as $10^{36} \, \text{m}^{-3}$. However, this is not the case due to the Pauli exclusion principle, which inhibit collisions. As a rule, the more collisional a system is, the larger is its associated coupling parameter. In the opposite way, extremely dense plasmas behave in an even more ideal way as the density increases because the occupation of the same quantum state by two spin

1/2 fermions is forbidden. Here, for the first time, we mention a quantum effect (the Pauli principle) having a rôle in plasma systems, namely yielding a smaller electron–electron collision frequency in comparison to a purely classical plasma. Moreover, since its beginning in the 1920s plasma physics has dealt basically with systems dilute enough so that quantum effects can be safely ignored. For this reason, plasma physics is generally considered to be almost classical. Sometimes quantum mechanics is necessary in limited instances of the usual plasma theory, as for the calculation of nuclear fusion cross sections. However, such are subsidiary quantum mechanical results, more related to nuclear physics than plasma physics, to be plugged in completely classical frameworks [22].

When are quantum effects relevant for plasmas? As mentioned, extremely dense plasmas behave like a quantum ideal gas, due to the exclusion principle. However, also dilute charged particle systems can exhibited quantum features, provided the dimensions of the system are small enough. Small enough here means dimensions comparable to the de Broglie wavelength

$$\lambda_{\mathrm{B}} = \frac{\hbar}{m v_{\mathrm{T}}}, \tag{1.1}$$

where $\hbar = h/(2\pi)$ is Planck's constant divided by 2π, m is the mass of the charge carriers and

$$v_{\mathrm{T}} = \left(\frac{\kappa_{\mathrm{B}} T}{m} \right)^{1/2} \tag{1.2}$$

is the thermal velocity. In this later expression, κ_{B} is Boltzmann's constant and T the thermodynamic temperature. Hence, if L_0 is a typical length scale of the plasma, quantum effects should be taken into account whenever $\lambda_{\mathrm{D}} \sim L_0$. This can be the case, for instance, in charged particle systems like semiconductor quantum wells, thin metal films, and nanoscale electronic devices in general [9, 18, 22]. If the de Broglie wavelength is comparable to the size of the system, the well-known quantum wave-like effects similar to those of physical optics (diffraction, interference, superposition) may take place in a decisive way. Such effects, for instance, show up in the resonant tunneling of electrons through a potential barrier in quantum devices, which in turn is decisive for the negative differential resistances of these systems [18, 26].

The de Broglie wavelength is a characteristic quantum length appropriate to weakly coupled systems where the charge carriers are not confined to a limited region of space. In the case of strongly coupled systems, able to form bound states as atoms, molecules and dense clusters, the de Broglie wavelength is replaced by the spatial extent of the particle's wavefunction [5].

Another relevant plasma parameter is the ambient magnetic field. Here, we have another instance where quantum effects can play a rôle, for instance, in the propagation of electron Bernstein modes in a degenerate plasma [11]. Moreover, the interconnection between spin dynamics, magnetic fields and ferromagnetic behavior has a pronounced influence [6,24], with importance, for example, in the propagation of spin oriented soliton structures in pair plasmas [7]. However, for simplicity in this chapter, magnetic fields are not included in the discussion.

One of the more basic collective properties of a plasma, be it classical or quantum, is a tendency for quasi-neutrality due to the shielding of any excess charge in the system. However, the way this shielding effect take place is different according to the classical or quantum nature of the plasma, as seen in the next section.

1.2   Debye Shielding in Degenerate and Nondegenerate Plasmas

Consider a test charge of magnitude $q_t > 0$ inserted into a plasma. For definiteness, assume the plasma to be composed by mobile electrons with number density n and a fixed ionic background with number density n_0. In a dynamic situation, the trajectories of the electrons would be deviated toward the test charge due to the Coulomb force. Therefore, a cloud of negative charge would form around the test charge, a phenomena known as shielding or screening. Hence, instead of the value q_t the test charge would acquire a smaller, effective shielded charge as seen from an observer outside the electron cloud. Assuming a quasi-equilibrium situation so that the electron number density $n = n(\mathbf{r})$ is a function of position only, the electrostatic field $\phi = \phi(\mathbf{r})$ is described by Poisson's equation

$$\nabla^2 \phi = \frac{e}{\varepsilon_0}(n(\mathbf{r}) - n_0) - \frac{q_t}{\varepsilon_0}\delta(\mathbf{r}). \tag{1.3}$$

Here $-e$ is the electron charge, ε_0 is vacuum's permittivity, and the test charge was conveniently placed at the origin. Moreover, the test charge was taken as sufficiently massive so that it can be regarded as motionless.

In this whole quiescent, quasi-static context, it is reasonable to assume that the electrons are in a local thermodynamic equilibrium. However, the nature of the equilibrium depends on the quantum or classical statistics obeyed by the electron gas. For sufficiently dilute systems [29,32], the indistinguishability between electrons need not to be taken into account. In this case, a local Maxwell–Boltzmann equilibrium can be assumed, so that

$$n = n_0 \exp\left(\frac{e\phi}{\kappa_B T}\right), \tag{1.4}$$

Actually, the number density comes from the one-particle distribution function $f(\mathbf{r},\mathbf{v})$ defined in such a way that $(1/N)f(\mathbf{r},\mathbf{v})\,d\mathbf{r}\,d\mathbf{v}$ gives the probability of finding one electron in an element $d\mathbf{r}\,d\mathbf{v}$ centered at position $\mathbf{r}$ with velocity $\mathbf{v}$. The normalization factor $1/N$, where N is the total number of electrons, is chosen so that

$$n = \int d\mathbf{v}\, f(\mathbf{r},\mathbf{v}). \tag{1.5}$$

Equation (1.4) is then consistent with the local Maxwell–Boltzmann equilibrium

$$f = n_0 \left(\frac{m}{2\pi \kappa_B T} \right)^{3/2} \exp\left[-\frac{1}{\kappa_B T} \left(\frac{mv^2}{2} - e\phi \right) \right]. \tag{1.6}$$

Even if collisions are not being explicitly discussed, it is reasonable, although not rigorous, to assume that collisions eventually produce the equilibrium (1.6).

We can suppose that before the introduction of the test charge, the ambient electrostatic field was zero. In this case, immediately after inserting q_t a sufficiently small scalar potential can be assumed, in a first approximation. In this case, we can Taylor expand (1.4) to first-order in the perturbation ϕ and substitute in (1.3) to get

$$\nabla^2 \phi = \frac{n_0 e^2}{\varepsilon_0 \kappa_B T} \phi - \frac{q_t}{\varepsilon_0} \delta(\mathbf{r}). \tag{1.7}$$

As discussed in textbooks [28], the solution to (1.7) with appropriate boundary conditions is the screened Coulomb potential

$$\phi = \frac{q_t}{4\pi \varepsilon_0 r} e^{-r/\lambda_D}, \tag{1.8}$$

where

$$\lambda_D = \left(\frac{\varepsilon_0 \kappa_B T}{n_0 e^2} \right)^{1/2} \tag{1.9}$$

is the electron Debye length. As apparent from (1.8), for distances larger than the Debye length the potential is very small. Hence, for all practical purposes λ_D is the effective range of the Coulomb interaction in a classical, Maxwellian plasma. Moreover, from the argument, it is evident that the shielding is a collective effect due to a large number of particles surrounding the test charge.

Now that we have a reasonable understanding about the shielding process in a dilute plasma, we can ask what happens in a nondilute, dense plasma. It follows from elementary quantum mechanics that we are not allowed to put all electrons in the same quantum state, since they are fermions satisfying Pauli's exclusion principle. An intuitive approach to this problem would assume an homogeneous one-particle probability distribution function in a sphere of radius v_F in velocity space, where v_F is the Fermi velocity [29, 32]. Therefore,

$$f = \frac{3n_0}{4\pi v_F^3} \quad \text{if} \quad \frac{mv^2}{2} - e\phi < E_F \tag{1.10}$$

and $f = 0$ otherwise. In (1.10), the Fermi energy $E_F = mv_F^2/2$ was introduced. Shortly, we will provide a recipe for expressing the Fermi energy as a function of the density. Moreover, in (1.10) the energy was shifted to allow for a nonzero scalar potential, in the same spirit of the energy-shifted Maxwell–Boltzmann equilibrium in (1.6).

The Thomas–Fermi [14] equilibrium (1.10) shows equal occupation probabilities for energies smaller than Fermi's energy, and zero occupation probabilities beyond. It is equivalent to a zero-temperature, local Fermi–Dirac distribution function [29, 32]. In addition, implicitly if we are using the statistical mechanical language of one-particle distribution functions, we are somehow neglecting quantum diffraction effects. Indeed, as will be seen in the next chapter, if we replace the (classical) one-particle distribution function by the equivalent quantum mechanical tool, the Wigner function, one cannot strictly assign occupation probabilities in phase space, since the Wigner function can assume negative values. This point will be discussed at length in Chap. 2.

With the above remarks, we can use (1.5) and the Thomas–Fermi equilibrium (1.10) to compute the number density. The result is

$$n = \int d\mathbf{v} f = n_0 \left(1 + \frac{e\phi}{E_F}\right)^{3/2}. \tag{1.11}$$

Poisson's equation as well as the Maxwell equations in general are valid for either classical or quantum systems. Hence, (1.11) can be linearized for small perturbation fields after inserted in Poisson's equation, to give

$$\nabla^2 \phi = \frac{3n_0 e^2}{2\varepsilon_0 E_F} \phi - \frac{q_t}{\varepsilon_0} \delta(\mathbf{r}). \tag{1.12}$$

Comparison of (1.7) and (1.12) shows that the classical screening effect still holds in the quantum (degenerate) case, but with the replacement $\kappa_B T \to E_F$, apart from a numerical factor of order 1. Correspondingly, the quantum screening distance, or Thomas–Fermi length λ_F, can be defined [36] as

$$\lambda_F = \left(\frac{2\varepsilon_0 E_F}{3n_0 e^2}\right)^{1/2}. \tag{1.13}$$

Notice that the thermodynamic temperature was assumed to be zero, so that $\lambda_D = 0$ while $\lambda_F \neq 0$. This is because in a zero-temperature classical electron gas the electron cloud around the test charge will have effectively zero radius. In the quantum realm, however, due to the exclusion principle the same electron cloud cannot collapse to a point in space (except with an infinite velocities dispersion which is absurd in view of the $T = 0$ assumption).

In an alternative version, each particle in the system can be interpreted as the test charge. Hence, the shielding effect hold for all particles, a manifestation of the charge neutrality tendency. Instead of its actual charge, for the Coulomb force what remains is a residual, screened charge. Moreover, for sufficiently large distances a charged particle in a plasma becomes invisible to the remaining of the system. In this context, notice that for the concept of screening length to be reasonable we need either $L_0 \gg \lambda_D$ or $L_0 \gg \lambda_F$ in the classical and quantum cases, respectively, where L_0 is the characteristic size of the system. Otherwise, each charge would be able to be "seen" by all remaining charges.

We have talked about Fermi energy and Fermi velocity. How to assign them a definite physical meaning? As described in statistical mechanics textbooks [29, 32], the question is how to accommodate a certain number N of fermions in a region of volume V. It is not allowed to put all fermions in the ground state, since at most two electrons can occupy the same orbital because of the spin number $1/2$. In this manner, the higher energy levels are filled up until the highest energy level is reached. The corresponding energy quantum number is the Fermi energy E_F. We omit here the details, but it can be shown [29, 32] that

$$E_F = \frac{\hbar^2}{2m}(3\pi^2 n_0)^{2/3}, \quad n_0 = N/V. \tag{1.14}$$

It is clear that for a sufficiently dilute electron gas $E_F \to 0$. Other quantities of interest are the Fermi temperature $T_F = E_F/\kappa_B$, the Fermi wavenumber $k_F = mv_F/\hbar$, and the Fermi momentum $\hbar k_F$.

1.3 Plasma Frequency

In the treatment of screening no temporal scales were involved. A dynamical, nonequilibrium situation arises if an electron charge depletion appears in the plasma. In this case, the Coulomb force tend to restore the charge neutrality. However, the displaced electrons are pushed back but due to the inertia they did not stay in the original equilibrium positions. Hence, a new charge depletion form, which again tend to be filled up by the restoring electric field. This is the mechanism for plasma oscillations, as described in textbooks [12, 28]. The characteristic time scale for plasma oscillations is given by ω_p^{-1}, where

$$\omega_p = \left(\frac{n_0 e^2}{m\varepsilon_0}\right)^{1/2} \tag{1.15}$$

is the plasma frequency. From the plasma frequency and the classical and quantum screening lengths, we obtain

$$\omega_p \lambda_D = \left(\frac{\kappa_B T}{m}\right)^{1/2} = v_T, \tag{1.16}$$

$$\omega_p \lambda_F = \left(\frac{2E_F}{3m}\right)^{1/2} = v_F/\sqrt{3}. \tag{1.17}$$

1.4   Energy Coupling Parameter

Classical weakly coupled plasmas tend to be hot and tenuous, while classical strongly coupled plasmas tend to be cold and dense. On the other hand, quantum plasmas are weakly coupled for sufficiently large densities, irrespective of the temperatures. Indeed, a many-body system will behave in a more ideal way provided its average potential energy is far less than its average kinetic energy. For both classical and quantum Coulomb systems, the mean interaction energy U_{pot} of one particle is

$$U_{\text{pot}} \sim \frac{e^2 n_0^{1/3}}{\varepsilon_0}, \tag{1.18}$$

since the mean inter-particle distance scales as $n_0^{-1/3}$ and taking into account only the electrostatic field. Hence, U_{pot} increases with the density, as expected.

However, the mean kinetic energies are different for classical and quantum systems. For a nondegenerate gas, one has [29, 32] the average kinetic energy K_C of a particle to be proportional to the temperature. On the other hand, in the degenerate case, one has the typical kinetic energy K_Q of a particle to be of the order of the Fermi energy, which increases with the density. Indeed, the more electrons are accommodated in a fixed region of space, the more their wavefunctions would significantly overlap, implying the enhancement of the Pauli pressure due to the exclusion principle. This increases the Fermi energy, as can be seen also from (1.14).

Hence, we can define the classical Γ_C and quantum Γ_Q coupling parameters as:

$$\Gamma_C = \frac{U_{\text{pot}}}{K_C} = \frac{U_{\text{pot}}}{\kappa_B T} = \frac{e^2 n_0^{1/3}}{\varepsilon_0 \kappa_B T} = 2.1 \times 10^{-4} \times \frac{n_0^{1/3}}{T}, \tag{1.19}$$

$$\Gamma_Q = \frac{U_{\text{pot}}}{K_Q} = \frac{U_{\text{pot}}}{\kappa_B T_F} = \frac{2me^2}{(3\pi^2)^{2/3}\varepsilon_0 \hbar^2 n_0^{1/3}} = 5.0 \times 10^{10} n_0^{-1/3}. \tag{1.20}$$

From the above expressions, we verify that classical weakly coupled plasmas tend to be dilute and cold, while quantum weakly coupled plasmas tend to be dense. The numerical values are for SI units.

Using the definition of Debye and Thomas–Fermi lengths, it can be proven that a system will not be strongly coupled provided the number of particles in a Debye or Fermi sphere is large. By definition, a Debye (Fermi) sphere has a radius λ_D (λ_F).

A more detailed survey on the implications of the quantum weak coupling assumption $\Gamma_Q \ll 1$ will be postponed to Sect. 2.5, where the validity conditions of the mean field approximation for quantum plasmas are discussed.

A fundamental dimensionless parameter remain to be introduced. It is the degeneracy parameter $\chi = T_F/T$, according to which the Fermi–Dirac statistics is unavoidable or not [29, 32]. Using the de Broglie wavelength $\lambda_B = \hbar/(mv_T)$, one find

$$\chi = \frac{T_F}{T} = \frac{1}{2}(3\pi^2 n_0 \lambda_B^3)^{2/3} \ll 1 \tag{1.21}$$

as the condition according to which the Maxwell–Boltzmann statistics can be used. Therefore, if the mean inter-particle distance is of the order of the de Broglie wavelength, the Fermi–Dirac statistics becomes necessary.

We have identified the more relevant scales in a many-body electrostatic charged particle system. These are: a quantum length scale, which for weakly coupled systems is the de Broglie wavelength according to which quantum diffraction effects are important or not; energy scales, namely the typical kinetic energies ($\sim \kappa_B T$ for a nondegenerate system, $\sim \kappa_B T_F$ for a degenerate system) and the interaction energy $\sim q^2 n_0^{1/3} / \varepsilon_0$. Weakly coupled systems (the theme of this monograph) are characterized by a small interaction energy in comparison to the average kinetic energy; temperature scales, namely the thermodynamic and Fermi temperatures, respectively, T and T_F. With them we can form the degeneracy parameter T_F / T which needs to be small for the Maxwell–Boltzmann statistics to be applicable. Otherwise, the Fermi–Dirac statistics should be employed, since the essential charge carriers in plasmas (the electrons) are fermions. In a minimal setting, we can, therefore, conclude that for electrostatic plasmas the fundamental nondimensional parameters can all be formed from the number density (or densities in the case of many species) and the temperature (temperatures), besides the characteristic size of the system when treating with nanometric systems.

1.5 Kinetic and Fluid Descriptions

In this section, some different levels of approximation for plasma modeling are discussed. As stressed, for example, in [3], there is a large number of combinations and permutations of the various descriptions in plasma physics. For instance, collisional or collisionless kinetic theory, two-fluid equations, magnetohydrodynamics, direct solving of the Newton equations of motion and so on. Choosing one or another method depends very much on the nature of the problem at hand. Here, a definite route is followed, so that a brief introduction to some of the most important plasma models can be sketched. In this route, we go from theories aiming at a more detailed knowledge of the state of the system, to theories focusing only on averaged properties of the plasma. Namely, we discuss first collisionless kinetic theory, then two-fluid equation modeling, and then magnetohydrodynamics, from the classical physics perspective. In this way, a model hierarchy for quantum plasma systems becomes self-evident, once the analog quantum tools can be found, by comparison with the classical methods.

In a Newtonian setting, the most fundamental point of view is to follow the trajectory of each charge in the plasma, as influenced by the electromagnetic force due to the remaining charges. Clearly, such approach is prohibitive, both numerically and computationally, due to the very large number of Newton's equations, which are coupled and nonlinear. Hence, a statistical, or kinetic approach seems appropriate. One can introduce the 1-particle distribution $f(\mathbf{r}, \mathbf{v}, t)$ defined in such a way that $f(\mathbf{r}, \mathbf{v}, t) \, \mathrm{d}\mathbf{r} \, \mathrm{d}\mathbf{v}$ gives the number of particles with position between $\mathbf{r}$ and $\mathbf{r} + \mathrm{d}\mathbf{r}$ and velocity between $\mathbf{v}$ and $\mathbf{v} + \mathrm{d}\mathbf{v}$ at the time t. Ignoring collisions and relativistic effects [3, 12, 28], the evolution equation solved by f is the Vlasov equation,

$$\frac{\partial f}{\partial t} + \mathbf{v} \cdot \frac{\partial f}{\partial \mathbf{r}} - \frac{e}{m} (\mathbf{E} + \mathbf{v} \times \mathbf{B}) \cdot \frac{\partial f}{\partial \mathbf{v}} = 0. \tag{1.22}$$

For definiteness, it was assumed a plasma composed by electrons (charge $-e$, mass m) in an immobile neutralizing ionic background. The electric field $\mathbf{E}$ and magnetic field $\mathbf{B}$ need to be found self-consistently solving the Maxwell equations, where the charge and current densities are found from f. In this way, one arrives at the Vlasov–Maxwell system, determining $f, \mathbf{E}$ and $\mathbf{B}$ under appropriate initial and boundary conditions. For instance, in the electrostatic case, the Maxwell equations reduce to Poisson's equation for the scalar potential ϕ,

$$\nabla^2 \phi = \frac{e}{\varepsilon_0} \left(\int \mathrm{d}\mathbf{v} \, f - n_0 \right). \tag{1.23}$$

Equations (1.22) and (1.23) constitute the Vlasov–Poisson system, to be supplemented by initial and boundary conditions.

If collisions need to be taken into account, some appropriate collision operator is inserted in the right-hand side of (1.22). Such collision operators may take for instance Lenard–Balescu [2, 20], Fokker–Planck [13, 31] or Bhatnagar–Gross–Krook [4] forms. However, collisional effects are outside the scope of the present book.

In the above statistical mechanics setting, we have a less detailed knowledge than following each particle's trajectory. Rather, we keep track of the flow of a bunch of particles starting in a small phase-space region centered at some initial position $\mathbf{r}$ and velocity $\mathbf{v}$. Nevertheless, the kinetic theory formulated in terms of the Vlasov–Maxwell system is frequently too complicated and some simplification advisable. Moreover the one-particle distribution function f often provide more knowledge than what is really needed. In this way, it is helpful to consider the moments, or weighted averages of f. By definition, an kth-order moment $M^{(k)}_{i...j}(\mathbf{r}, t)$ of f is given by the integral

$$M^{(k)}_{i...j}(\mathbf{r}, t) = \int \mathrm{d}\mathbf{v} \, v_i \dots v_j f(\mathbf{r}, \mathbf{v}, t), \tag{1.24}$$

where the number of velocity components in the right-hand side of (1.24) is k.

For instance, one can consider the zeroth-order moment given by (1.5), which yields the number density $n(\mathbf{r}, t)$, and correspondingly the charge density $-en(\mathbf{r}, t)$. Moreover, since f is proportional to the probability to find a random particle at a point $\mathbf{r}$ and with velocity $\mathbf{v}$, a global velocity $\mathbf{u}(\mathbf{r}, t)$ can be introduced so that

$$n(\mathbf{r}, t) \mathbf{u}(\mathbf{r}, t) = \int \mathrm{d}\mathbf{v} \, \mathbf{v} \, f(\mathbf{r}, \mathbf{v}, t). \tag{1.25}$$

Hence, the first-order moment of f fives the global current density $n(\mathbf{r},t)\mathbf{u}(\mathbf{r},t)$ and correspondingly the global electric current density $-en(\mathbf{r},t)\mathbf{u}(\mathbf{r},t)$. Similarly from the higher-order moments, one can obtain information about the energy density and the heat transport vector of the plasma [28].

One can be tempted to study the evolution equations solved by the moments. In this way, we can find a set of macroscopic, or fluid equations. However, the equation of continuity solved by the zeroth-order moment, the number density n, involves the fluid velocity $\mathbf{u}$, which is a higher-order moment. In the same way, the momentum transport equation solved by $\mathbf{u}$ contain a higher-order moment, namely the pressure dyad. In general, the equation satisfied by the kth-order moment contains the $(k+1)$th-order moment, so that an infinite chain of equations is found [3, 12, 28]. Hence, one is faced with a closure problem, frequently solved by means of some ad hoc quasi-equilibrium hypothesis, which is equivalent to postulate an appropriate equation of state. Underlying the closure assumption, there is some knowledge about the phenomena under analysis. For instance, for fast processes, an adiabatic equation of state can be used to specify the form of the scalar pressure. On the other hand, for slow processes allowing for thermalization, an isothermal equation of state can be used to define the scalar pressure entering the momentum transport equation. The choice of equation of state will be detailed in Sect. 4.4.

Allowing for mobile ions or other species in general, there will be a reduced one-particle distribution function for each specie. This in turn would imply one kinetic (Vlasov) equation for each specie. Repeating then the average procedure, in the case of a two-species, or electron–ion plasma, we would obtain a two-fluid model. Implicitly, in this model charge separation is relevant enough so as to deserve some attention. However, for high conductivity plasmas and slow-time scale phenomena, one can somehow merge the electronic and ionic fluids, appropriately assigning global charge and current densities. In this way, one find the magnetohydrodynamic modeling for plasmas, which is the less detailed description among all discussed in this section. Nevertheless, in classical plasma physics, this is one of the most useful models, applied to a huge variety of laboratory and natural plasmas [3, 12, 28]. We postpone a more proper account on magnetohydrodynamic theory to Chap. 6.

In summary, we have commented on the following plasma physics descriptions, going from the more detailed to the more simplified approaches: collisionless kinetic (Vlasov–Maxwell system) $\rightarrow$ macroscopic multi-fluid (or one fluid in the case of an electron gas in a fixed ionic background) $\rightarrow$ magnetohydrodynamics. While obviously this is not the only possible route trough the jungle of plasma models, it provides a logic perspective to be followed also in the quantum realm. Indeed, in spite of the many neglected phenomena, the fluid approaches have the merit of simplicity in comparison to kinetic theory. The nonlinear aspects of quantum plasmas can hardly be accessed with microscopic, not macroscopic settings, except perhaps through expensive numerical or painful analytic calculations. Thanks to the quantum hydrodynamic approach, besides practical advances in the manipulation of nanoscale systems or in laser ablation experiences, we observe today a fast development in quantum plasma physics.

1.6   Historical Notes

The most immediate example of a quantum plasma is provided by the quantum electron gas in a metal. For this reason, the analysis of the ground state and correlation energies of an electron gas has been the subject of many works in the 1950s, in particular for high densities where the Fermi–Dirac statistics is necessary [10, 15, 33]. This approach usually considered the application of quantum field-theoretic techniques in many-body perturbation theory.

On the other hand, dynamic properties have been investigated by the pioneers in view of the propagation of linear waves in dense plasmas. The main tool in this approach was the self-consistent, kinetic modeling in terms of the collisionless quantum Boltzmann equation [19, 21, 30].

The main drawback of the field-theoretic and/or kinetic treatments is the analytic complexity. Only recently, with the introduction of macroscopic theories [16,17,23], the interest on quantum plasmas has gained a new impulse, see [22, 36] for reviews.

We also remark on two of the main basically *missing* topics of this monograph: relativistic and spin quantum plasmas. Quantum plasmas where the spin dynamics is taken into account not only through quasi-equilibrium properties (e.g., assuming an equation of state appropriated for a degenerate electron gas) have been subject to intense research in the last years [6–8, 24, 35, 37]. More recently, the joint spin and relativistic effects have started to be incorporated in terms of macroscopic models [1,34], in contrast to kinetic approaches [27]. In an energy ordering, the macroscopic models for quantum plasmas have been already constructed: (a) in the nonrelativistic quantum mechanics context (tunneling, wave packet spreading); (b) in the lowest-order relativistic mechanics context (magnetization, spin–orbit coupling). To go for macroscopic quantum plasma models in the fully relativistic and quantum field theoretical aspects (*zitterbewegung*, pair production, etc.) is currently a challenge [25].

Here, we finish our brief introduction on the physics of classical and quantum plasmas. In the next chapter, we consider the Wigner–Poisson system, which is the natural kinetic model for collisionless quantum plasmas. We discuss some applications as well as the validity conditions for such modeling.

Problems

1.1. Consult the literature, find typical temperatures and densities and classify the following systems as degenerate/nondegenerate and strongly/weakly coupled: semiconductor quantum well; gas discharge; solar corona; thin metal film; white dwarf; electron gas in metals.

1.2. Repeat the last problem checking which of the listed charged particle systems can be taken as nonrelativistic, computing v_T/c and v_F/c, where v_T and v_F are the thermal and Fermi velocities and c is the speed of light.

1.3. Use (1.4) and (1.8) to check the global charge neutrality to first-order in the interaction strength. In other words, show that $e \int (n - n_0) d\mathbf{r} = q_t$ to first-order.

1.4. Show that a system will be *not* strongly coupled provided the number of particles in a Debye or Fermi sphere is large, according to the classical or quantum nature of it.

References

1. Asenjo, F. A., Munoz, V., Valdivia, J. A. and Mahajan, S. M.: A hydrodynamical model for relativistic spin quantum plasmas. Phys. Plasmas **18**, 012107–012118 (2011)
2. Balescu, R.: Irreversible processes in ionized gases Phys. Fluids **3**, 52–63 (1960)
3. Belan, P. M.: Fundamentals of Plasma Physics. Cambridge, New York (2006)
4. Bhatnagar, P. L., Gross, E. P. and Krook, M.: A model for collision processes in gases. I. Small amplitude processes in charged and neutral one-component systems. Phys. Rev. **94**, 511–525 (1954)
5. Bonitz, M., Semkat, D., Filinov, A., Golubnychyi, V., Kremp, D., Gericke, D. O., Murillo, M. S., Filinov, V., Fortov, V., Hoyer, W. and Koch, S. W.: Theory and simulation of strong correlations in quantum Coulomb systems. J. Phys. A: Math. Gen. **36**, 5921–5930 (2003)
6. Brodin, G. and Marklund, M.: Spin magnetohydrodynamics. New J. Phys. **9**, 277–288 (2007)
7. Brodin, G. and Marklund, M.: Spin solitons in magnetized pair plasmas. Phys. Plasmas **14**, 1121071–1121075 (2007)
8. Brodin, G., Misra, A. P. and Marklund, M.: Spin contribution to the ponderomotive force in a plasma. Phys. Rev. Lett. **105**, 105004–105008 (2010)
9. Di Ventra, M.: Electrical Transport in Nanoscale Systems. Cambridge, New York (2008)
10. Dubois, D. F.: Electron interactions I - field theory of a degenerate electron gas. Ann. Phys. **7**, 174–237 (1959)
11. Eliasson, B. and Shukla, P. K.: Numerical and theoretical study of Bernstein modes in a magnetized quantum plasma. Phys. Plasmas **15**, 1021011–1021025 (2008)
12. Fitzpatrick, R.: The Physics of Plasmas. Lulu Inc., Raleigh (2011)
13. Fokker, A. D.: Die mittlere energie rotierender elektrischer dipole im strahlungsfeld, Ann. Phys. **348**, 810–820 (1914)
14. Frensley, W. R.: Boundary conditions for open systems driven far from equilibrium. Rev. Mod. Phys. **62**, 745–791 (1990)
15. Gellmann, M. and Brueckner, K. A.: Correlation energy of an electron gas at high density. Phys. Rev. **106**, 364–368 (1957)
16. Haas, F., Manfredi, G., Feix, M.: Multistream model for quantum plasmas. Phys. Rev. E **62**, 2763–2772 (2000)
17. Haas, F.: A magnetohydrodynamic model for quantum plasmas. Phys. Plasmas **12**, 062117–062126 (2005)
18. Jüngel, A.: Transport Equations for Semiconductor Devices. Springer, Berlin-Heidelberg (2009)
19. Klimontovich, Y. and Silin, V. P.: The spectra of systems of interacting particles. In: Drummond, J. E. (ed.) Plasma Physics, pp. 35–87, McGraw-Hill, New York (1961)
20. Lenard, A.: On Bogoliubov's kinetic equation for a spatially homogeneous plasma. Ann. Phys. **10**, 390–400 (1960)
21. Lindhard, J.: On the properties of a gas of charged particles. Dan. Vidensk. Selsk., Mat. Fys. Medd. **28**, 1–57 (1954)
22. Manfredi, G.: How to model quantum plasmas. Fields Inst. Commun. **46**, 263–287 (2005)

23. Manfredi, G., Haas, F.: Self-consistent fluid model for a quantum electron gas. Phys. Rev. B **64**, 075316–075323 (2001)
24. Marklund, M. and Brodin, G.: Dynamics of spin 1/2 quantum plasmas. Phys. Rev. Lett. **98**, 025001–025005 (2007)
25. Marklund, M. and Lundin, J.: Quantum vacuum experiments using high intensity lasers. Eur. J. Phys. D **55**, 319–326 (2009)
26. Markowich, P. A., Ringhofer, C. A., Schmeiser, C.: Semiconductor Equations. Springer, Wien (1990)
27. Melrose, D. B.: Quantum Plasmadynamics: Unmagnetized Plasmas, Lecture Notes in Physics Vol. 735. Springer, New York (2008)
28. Nicholson, D. R.: Introduction to Plasma Theory. John Wiley, New York (1983)
29. Pathria, R.K.: Statistical Mechanics, 2nd ed. Butterworth-Heinemann, Woburn (1996)
30. Pines, D. and Nozières, P. The Theory of Quantum Liquids. New York, W. A. Benjamin (1966)
31. Planck, M.: Über einen satz der statistischen dynamik und seine erweiterung in der quantentheorie. Sitzber. Preuss. Akad. Wiss., Phys-Math. Klasse, 324–341 (1917)
32. Salinas, S. R. A.: Introduction to Statistical Physics. Springer, New York (2001)
33. Sawada, K.: Correlation energy of an electron gas at high density. Phys. Rev. **106**, 372–383 (1957)
34. Shukla, P. K. and Eliasson, B.: Nonlinear interactions between electromagnetic waves and electron plasma oscillations in quantum plasma. Phys. Rev. Lett. **99**, 096401–096405 (2007)
35. Shukla, P. K.: A new spin on quantum plasmas. Nature Phys. **5**, 92–93 (2009)
36. Shukla, P. K. and Eliasson, B.: Nonlinear aspects of quantum plasma physics. Phys. Uspekhi **53**, 51–76 (2010)
37. Zamanian, J. Marklund, M. and Brodin, G.: Scalar quantum kinetic theory for spin-1/2 particles: mean field theory. New J. Phys. **12**, 043019–043048 (2010)

Chapter 2
The Wigner–Poisson System

Abstract In electrostatic quantum plasmas, the Wigner–Poisson system plays the same rôle as the Vlasov–Poisson system in classical plasmas. This chapter considers the basic properties of the Wigner–Poisson system, including the essentials on the Wigner function method and the derivation of the Wigner–Poisson system in the context of a mean field theory. This chapter also contains a discussion on the Schrödinger–Poisson system as well as extensions to include correlation and collisional effects. The Wigner–Poisson system is shown to imply, in the high-frequency limit, the Bohm–Pines dispersion relation for linear waves, which is the quantum analog of the Bohm–Gross dispersion relation for classical plasmas.

2.1 The Wigner Function

To maintain the closest resemblance to the familiar methods of classical plasma physics, the Wigner function approach is the natural choice. Indeed, using the Wigner function, one can proceed in almost total analogy with the standard phase-space distribution function method to compute macroscopic quantities like number and current densities. Hence, it is useful to review some of the properties of the Wigner (pseudo) distribution function approach. In addition, the differences between classical and quantum formalisms will be highlighted. The treatment is by no means exhaustive, being intentionally restricted to the bare necessary minimum. More complete reviews on Wigner function methods can be found, for example, in [8, 18, 26, 36].

For simplicity, let us start with a one-dimensional, one-particle pure state quantum system, represented by a wavefunction $\psi(x,t)$. In this case, the Wigner function $f = f(x, v, t)$ is defined [38] as

$$f = \frac{m}{2\pi\hbar} \int ds \, \exp\left(\frac{imvs}{\hbar}\right) \psi^*\left(x + \frac{s}{2}, t\right) \psi\left(x - \frac{s}{2}, t\right), \qquad (2.1)$$

F. Haas, *Quantum Plasmas: An Hydrodynamic Approach*, Springer Series on Atomic, Optical, and Plasma Physics 65, DOI 10.1007/978-1-4419-8201-8_2,
© Springer Science+Business Media, LLC 2011

where x is the position, v the velocity, t the time, m the particle's mass and $\hbar = h/(2\pi)$, where h is Planck's constant. In the above equation, the integration limits goes from minus to plus infinity, a convention followed except otherwise stated. The Wigner function provides a phase-space description of the quantum system where all physical quantities can be found from the kth-order moments $\int dv\, v^k f(x,v,t)$. For instance, both the probability density

$$n(x,t) = |\psi(x,t)|^2 = \int dv\, f(x,v,t) \tag{2.2}$$

and the probability current

$$J(x,t) = \frac{i\hbar}{2m}\left(\psi\frac{\partial \psi^*}{\partial x} - \psi^*\frac{\partial \psi}{\partial x}\right) = \int dv\, v\, f(x,v,t) \tag{2.3}$$

can be readily obtained, respectively, from the zeroth and first order moments of the Wigner function. Here, we assume ψ normalized to unity. Moreover, the Wigner function is always real, differently from the wavefunction which is complex.

In the more general case of a mixed state, the one-dimensional, one-particle system is represented by a quantum statistical mixture $\{\psi_\alpha(x,t), p_\alpha\}, \alpha = 1,2,\ldots,M$, where each wavefunction $\psi_\alpha(x,t)$ occurs with a probability p_α such that $p_\alpha \geq 0$, $\sum_{\alpha=1}^M p_\alpha = 1$. In such a situation, the Wigner function is given by the superposition

$$f = \frac{m}{2\pi\hbar}\sum_{\alpha=1}^M p_\alpha \int ds\, \exp\left(\frac{imvs}{\hbar}\right)\psi_\alpha^*\left(x+\frac{s}{2},t\right)\psi_\alpha\left(x-\frac{s}{2},t\right). \tag{2.4}$$

The corresponding generalization of (2.2) and (2.3) is then

$$n(x,t) = \sum_{\alpha=1}^M p_\alpha |\psi_\alpha(x,t)|^2 = \int dv\, f(x,v,t), \tag{2.5}$$

$$J(x,t) = \frac{i\hbar}{2m}\sum_{\alpha=1}^M p_\alpha\left(\psi_\alpha\frac{\partial \psi_\alpha^*}{\partial x} - \psi_\alpha^*\frac{\partial \psi_\alpha}{\partial x}\right) = \int dv\, v\, f(x,v,t). \tag{2.6}$$

Notice that the quantum fluid probability $n(x,t)$ and the current density $J(x,t)$ are still given in terms of the zeroth- and first-order moments of $f(x,v,t)$.

The Wigner formalism can be rephrased in the density matrix $\rho(x,y,t)$ language, since

$$\rho(x,y,t) \equiv \sum_{\alpha=1}^M p_\alpha \psi_\alpha(x,t)\psi_\alpha^*(y,t) = \int dv\, \exp\left(\frac{imv(x-y)}{\hbar}\right) f\left(\frac{x+y}{2},v,t\right), \tag{2.7}$$

with the inverse transformation being

$$f(x,v,t) = \frac{m}{2\pi\hbar} \int ds \, \exp\left(\frac{imvs}{\hbar}\right) \rho\left(x+\frac{s}{2}, x-\frac{s}{2}, t\right). \tag{2.8}$$

In the sense of the equivalence implied by (2.7) and (2.8), the use of the Wigner function or the density matrix is just a question of taste.

From the Wigner function, we can readily derive the marginal probability distributions in coordinate and momentum space. Indeed, from (2.8), it follows that

$$\int dv f = \rho(x,x,t) \tag{2.9}$$

and

$$\int dx f = m\tilde{\rho}(p,p',t), \tag{2.10}$$

where

$$\tilde{\rho}(p,p',t) = \frac{1}{2\pi\hbar} \int dx dx' \, \exp\left(\frac{i}{\hbar}(p'x' - px)\right) \rho(x,x',t) \tag{2.11}$$

denotes the matrix components of the density operator in the momentum representation, with $p = mv, p' = mv'$.

Going one step further, consider now a N-particle statistical mixture described by the set $\{\psi_\alpha^N(x_1, x_2, \ldots, x_N, t), p_\alpha\}$, where the normalized N-particle ensemble wavefunctions $\psi_\alpha^N(x_1, x_2, \ldots, x_N, t)$ are distributed with probabilities $p_\alpha, \alpha = 1, 2, \ldots, M$ satisfying $p_\alpha \geq 0, \sum_{\alpha=1}^M p_\alpha = 1$ as before. Here, x_i represents the position of the ith-particle, $i = 1, 2, \ldots, N$. For simplicity, assume all particles to have the same mass m. In analogy with (2.4) the N-particle Wigner function is then defined as

$$f^N(x_1, v_1, \ldots, x_N, v_N, t) = N \left(\frac{m}{2\pi\hbar}\right)^N \sum_{\alpha=1}^M p_\alpha \int ds_1, \ldots, ds_N \, \exp\left(\frac{im\sum_{i=1}^N v_i s_i}{\hbar}\right)$$

$$\times \psi_\alpha^{N*}\left(x_1 + \frac{s_1}{2}, \ldots, x_N + \frac{s_N}{2}, t\right)$$

$$\times \psi_\alpha^N\left(x_1 - \frac{s_1}{2}, \ldots, x_N - \frac{s_N}{2}, t\right). \tag{2.12}$$

The factor N in (2.12) is inserted so that

$$\int dx_1 \, dv_1, \ldots, dx_N \, dv_N \, f^N(x_1, v_1, \ldots, x_N, v_N, t) = N. \tag{2.13}$$

In this manner, the integral of f^N over all the velocities gives a number density. For simplicity, at this stage we are not taking into account the usually fermionic character of the charged particles in plasma, so that the N-body ensemble wavefunctions are not necessarily antisymmetrized.

From the N-particle Wigner function, the expectation value of any observable can be computed in the same way as in classical statistical mechanics. In other words, $f(x_1, v_1, \ldots, x_N, v_N, t)$ act as a weight in the same sense of the classical N-body particle distribution function $f_{cl}^N(x_1, v_1, \ldots, x_N, v_N, t)$. In this context, we have that $(1/N) f_{cl}^N(x_1, v_1, \ldots, x_N, v_N, t) dx_1 dv_1, \ldots, dx_N dv_N$ gives the probability of the particle 1 being in an area $dx_1 dv_1$ in phase space centered at position x_1 and velocity v_1, the particle 2 being in an area $dx_2 dv_2$ centered at position x_2 and velocity v_2 and so on. However, the Wigner function is not positive definite, so that it is not a probability but a pseudo-probability distribution.

More exactly, suppose a classical phase-space function $A(x_1, v_1, \ldots, x_N, v_N, t)$ corresponding to a self-adjoint quantum mechanical operator $\hat{A}(\hat{x}_1, \hat{v}_1, \ldots, \hat{x}_N, \hat{v}_N, t)$. Here, the hats denote operators and everything could be rewritten in terms of momenta and not velocities. We prefer to use velocities instead of momenta to assure a manifestly gauge invariant formalism. In addition, the transition from functions to operators is by no means unique: a well-defined correspondence rule should be employed [18, 26]. Special care should be paid with more complicated phase space observables involving noncommuting objects like products of functions of position and momenta. Indeed, to calculate expectation values using the Wigner formalism, we need first to map the observable into a phase-space function using the Weyl correspondence [37]. In practice, this is equivalent to ordering operators into a symmetric product of the position and momenta operators, using the commutation relations and then making the replacements $\hat{x}_i \to x_i$ and $\hat{p}_i \to p_i$, where $\hat{x}_i, \hat{p}_i$ are the position and momentum operators of the ith-particle and x_i, p_i the corresponding classical position and momentum functions.

Given a classical function $A(x_1, v_1, \ldots, x_N, v_N, t)$ the phase-space average $\langle A \rangle_{cl}$ is

$$\langle A \rangle_{cl} = \frac{1}{N} \int dx_1 dv_1, \ldots, dx_N dv_N f_{cl}^N(x_1, v_1, \ldots, x_N, v_N, t) A(x_1, v_1, \ldots, x_N, v_N, t),$$

(2.14)

while the expectation value $\langle \hat{A} \rangle$ of the associated Weyl ordered self-adjoint operator $\hat{A}(\hat{x}_1, \hat{v}_1, \ldots, \hat{x}_N, \hat{v}_N, t)$ is

$$\langle \hat{A} \rangle = \frac{1}{N} \int dx_1 dv_1, \ldots, dx_N dv_N f^N(x_1, v_1, \ldots, x_N, v_N, t) A(x_1, v_1, \ldots, x_N, v_N, t).$$

(2.15)

We have an explicit resemblance between classical and quantum formalisms.

As an example, one can be interested in the expectation value of the total energy of an interacting system with potential energy $V(x_1, \ldots, x_N)$. This average is given by

$$\frac{1}{N} \int dx_1 dv_1, \ldots, dx_N dv_N f^N(x_1, v_1, \ldots, x_N, v_N, t) \left(\sum_{i=1}^N \frac{mv_i^2}{2} + V(x_1, \ldots, x_N) \right),$$

(2.16)

which is the same expression used in classical statistical mechanics, with the replacement $f_{cl}^N \to f^N$.

As already remarked, to obtain nonerroneous expectation values of operators involving noncommuting observables using the Wigner formalism, the Weyl ordering should be employed. To see an illustrative example, consider the operator

$$\hat{x}_i \hat{p}_j = \frac{1}{2}(\hat{x}_i \hat{p}_j + \hat{p}_j \hat{x}_i) + \frac{1}{2}[\hat{x}_i, \hat{p}_j]$$

$$= \frac{1}{2}(\hat{x}_i \hat{p}_j + \hat{p}_j \hat{x}_i) + \frac{i\hbar \delta_{ij}}{2}$$

$$\rightarrow x_i p_j + \frac{i\hbar \delta_{ij}}{2} \quad \text{(Weyl rule)}, \tag{2.17}$$

using the commutation relation $[\hat{x}_i, \hat{p}_j] = i\hbar\delta_{ij}$ and where in the last equality the Weyl correspondence was applied. Hence, the required expectation value is

$$\langle \hat{x}_i \hat{p}_j \rangle = \frac{m}{N} \int dx_1 dv_1, \ldots, dx_N dv_N f^N(x_1, v_1, \ldots, x_N, v_N, t) x_i v_j + \frac{i\hbar \delta_{ij}}{2}, \tag{2.18}$$

taking into account $p_j = mv_j$. Indeed, after some integrations by parts the right-hand side of (2.18) is found to be

$$\frac{m}{N} \int dx_1 dv_1, \ldots, dx_N dv_N f^N(x_1, v_1, \ldots, x_N, v_N, t) x_i v_j + \frac{i\hbar \delta_{ij}}{2}$$

$$= -i\hbar \sum_{\alpha=1}^{M} p_\alpha \int dx_1, \ldots, dx_N \, \psi_\alpha^{N*}(x_1, \ldots, x_N, t) x_i \frac{\partial}{\partial x_j} \psi_\alpha^N(x_1, \ldots, x_N, t), \tag{2.19}$$

in line with the coordinate representation $\hat{p}_j \rightarrow -i\hbar\partial/\partial x_j$ of the momentum operator $\hat{p}_j$ (not to confound with the ensemble probabilities p_α).

In addition to the Wigner function, alternative quantum probability distribution functions can be constructed. Among them, we can cite at least the Glauber–Sudarshan function [11, 34], the Q-function [12], the Husimi function [19], the Kirkwood distribution function [22] and the standard-ordered distribution function [32]. For these alternative functions, the underlying quantum-classical correspondence is given by specific methods other than the Weyl rule [8, 26]. The relevance of the non-Wigner probability distributions is recognized for specific purposes. For instance, the Q-functions and the Husimi functions can be shown to be everywhere nonnegative [26]. However, the Wigner function has a number of simultaneous attractive properties which makes it more popular than the other distribution functions. For example, the Q and Husimi functions does not provide the marginal probability distributions in coordinate and momentum space in the sense of (2.9) and (2.10), besides not satisfying certain closure properties [8, 26].

We note that in some places in the literature the Q-function and Husimi's function are considered as synonymous, as a brief survey reveal. However, although their expressions are the same, their correspondence rules are not, see [26].

2.2 Mean Field Approximation

In analogy to classical statistical mechanics, it is useful to introduce the reduced one-particle Wigner function $f(x_1, v_1, t)$,

$$f(x_1, v_1, t) = \int dx_2 dv_2, \ldots, dx_N dv_N\, f^N(x_1, v_1, \ldots, x_N, v_N, t), \qquad (2.20)$$

the reduced two-particle Wigner function $f^{(2)}(x_1, v_1, x_2, v_2, t)$ with a convenient normalization factor N,

$$f^{(2)}(x_1, v_1, x_2, v_2, t) = N \int dx_3 dv_3, \ldots, dx_N dv_N\, f^N(x_1, v_1, \ldots, x_N, v_N, t), \qquad (2.21)$$

as well as the remaining reduced i-particle Wigner functions, $i = 3, \ldots, N$. If the Wigner function were a true probability distribution, $(1/N) f(x_1, v_1, t) dx_1 dv_1$ would give the probability of finding the particle 1 in an area $dx_1 dv_1$ centered at (x_1, v_1), irrespective of the "position" and "velocity" of the ith-particles, $i = 2, \ldots, N$. The other partial, or reduced Wigner functions would have a similar interpretation.

What is the evolution equation satisfied by the N-body Wigner function? To answer the question, we follow the philosophy of [24]. Let us start from the Schrödinger equation satisfied by the N-body ensemble wavefunctions,

$$i\hbar \frac{\partial \psi_\alpha^N}{\partial t} = -\frac{\hbar^2}{2m} \sum_{i=1}^{N} \frac{\partial^2 \psi_\alpha^N}{\partial x_i^2} + V(x_1, \ldots, x_N)\, \psi_\alpha^N \qquad (2.22)$$

for a potential energy $V(x_1, \ldots, x_N)$.

From (2.12) and (2.22),

$$\frac{\partial f^N}{\partial t} = N \left(\frac{m}{2\pi\hbar}\right)^N \sum_{\alpha=1}^{M} p_\alpha \int ds_1, \ldots, ds_N \exp\left(\frac{im \sum_{i=1}^{N} v_i s_i}{\hbar}\right)$$

$$\times \left[\frac{i\hbar}{2m} \sum_{j=1}^{N} \left(\psi_\alpha^{N*}(\mathbf{x}_+, t) \frac{\partial^2 \psi_\alpha^N(\mathbf{x}_-, t)}{\partial x_j^2} - \frac{\partial^2 \psi_\alpha^{N*}(\mathbf{x}_+, t)}{\partial x_j^2} \psi_\alpha^N(\mathbf{x}_-, t) \right) \right.$$

$$\left. + \frac{i}{\hbar} \left(V(\mathbf{x}_+) - V(\mathbf{x}_-) \right) \psi_\alpha^{N*}(\mathbf{x}_+, t) \psi_\alpha^N(\mathbf{x}_-, t) \right]. \qquad (2.23)$$

Above, we introduced the displaced collective coordinates

$$\mathbf{x}_+ = (x_1 + s_1/2, \ldots, x_N + s_N/2), \qquad (2.24)$$

$$\mathbf{x}_- = (x_1 - s_1/2, \ldots, x_N - s_N/2). \qquad (2.25)$$

To deal with (2.23) and similar equations, the following two identities are useful,

$$
\psi_\alpha^{N*}(\mathbf{x}_+,t)\frac{\partial^2 \psi_\alpha^N(\mathbf{x}_-,t)}{\partial x_i^2} - \frac{\partial^2 \psi_\alpha^{N*}(\mathbf{x}_+,t)}{\partial x_i^2}\psi_\alpha^N(\mathbf{x}_-,t)
$$

$$
= -2\frac{\partial^2}{\partial x_i \partial s_i}\left[\psi_\alpha^{N*}(\mathbf{x}_+,t)\,\psi_\alpha^N(\mathbf{x}_-,t)\right],
\tag{2.26}
$$

$$
N\sum_{\alpha=1}^{M} p_\alpha \psi_\alpha^{N*}(\mathbf{x}_+,t)\,\psi_\alpha^N(\mathbf{x}_-,t)
$$

$$
= \int dv_1,\dots,dv_N \exp\left(-\frac{im\sum_{i=1}^{N} v_i s_i}{\hbar}\right) f^N(x_1,v_1,\dots,x_N,v_N,t).
\tag{2.27}
$$

From these identities and after integration by parts, (2.23) is converted into

$$
\frac{\partial f^N}{\partial t} + \sum_{i=1}^{N} v_i \frac{\partial f^N}{\partial x_i}
$$

$$
= \int dv_1',\dots,dv_N' \times K^N[V\,|\,v_1'-v_1,x_1,\dots,v_N'-v_N,x_N,t]\,f^N(x_1,v_1',\dots,x_N,v_N',t),
\tag{2.28}
$$

introducing the functional

$$
K^N[V\,|\,v_1'-v_1,x_1,\dots,v_N'-v_N,x_N,t]
$$

$$
= -\frac{i}{\hbar}\left(\frac{m}{2\pi\hbar}\right)^N \int ds_1,\dots,ds_N \exp\left(-\frac{im\sum_{i=1}^{N}(v_i'-v_i)s_i}{\hbar}\right)(V(\mathbf{x}_+)-V(\mathbf{x}_-)).
\tag{2.29}
$$

In principle for a given interaction potential solving (2.28) amounts to a complete description of the N-body quantum problem. However, the development of analytic and numerical techniques for the N-body problem is of course a tremendous task. Moreover, for practical reasons, it is more effective to deal with the reduced Wigner functions, since f^N contain far more information than what is needed. In this regard, the one-particle Wigner function $f(x,v,t)$ plays a privileged rôle. Indeed, most macroscopic objects like number $n(x,t)$ and current $J(x,t)$ densities can be derived from the moments of $f(x,v,t)$,

$$
n(x,t) = \int dv\, f(x,v,t),
\tag{2.30}
$$

$$
J(x,t) = \int dv\, f(x,v,t)\,v,
\tag{2.31}
$$

much in the same way as in the classical formalism, in the transition from kinetic to fluid models. We also observe that (2.5) and (2.6) hold for a one-particle quantum fluid system, differently than (2.30) and (2.31) which apply to the N-body problem.

From the above reasoning, it is clearly relevant to obtain the evolution equation for the one-body reduced function $f(x,v,t)$. We are specially concerned with the case where the system components interact through some two-body potential W,

$$V(x_1,\ldots,x_N) = \sum_{i<j} W(|x_i - x_j|). \tag{2.32}$$

The importance of such a situation is evident due to the Coulomb forces present in charged particle systems.

Integrating (2.28) in the $(x_2,v_2,\ldots,x_N,v_N)$ variables, it follows that

$$\frac{\partial f}{\partial t} + v_1 \frac{\partial f}{\partial x_1} = -\frac{im}{2\pi\hbar^2} \int ds_1 \, dx_2 \ldots, dx_N \, dv_1',\ldots, dv_N' \exp\left(-\frac{im(v_1' - v_1)s_1}{\hbar}\right)$$

$$\times \sum_{i=1}^{N} \left(W\left(\left|x_1 - x_i + \frac{s_1}{2}\right|\right) - W\left(\left|x_1 - x_i - \frac{s_1}{2}\right|\right)\right)$$

$$\times f^N(x_1,v_1',\ldots,x_N,v_N',t). \tag{2.33}$$

A change of variables shows that the latter can be written as

$$\frac{\partial f}{\partial t} + v_1 \frac{\partial f}{\partial x_1} = -\frac{im}{2\pi\hbar^2} \int ds_1 \, dv_1' \, dx_2 \, dv_2' \exp\left(-\frac{im(v_1' - v_1)s_1}{\hbar}\right)$$

$$\times \left(W\left(\left|x_1 - x_2 + \frac{s_1}{2}\right|\right) - W\left(\left|x_1 - x_2 - \frac{s_1}{2}\right|\right)\right)$$

$$\times f^{(2)}(x_1,v_1',x_2,v_2',t). \tag{2.34}$$

in terms of the reduced two-particle Wigner function $f^{(2)}$ defined in (2.21). In the derivation $N \gg 1$ was taken into account. Actually, a more detailed argument involving the higher-order Wigner functions yield a quantum BBGKY (Bogoliubov–Born–Green–Kirkwood–Yvon) hierarchy [3, 6, 23, 39], where the dynamics of the $(N-1)$-body reduced Wigner function is shown to depend on the N-body reduced Wigner function. Hence, in both the classical infinite BBGKY set of equations and its quantum analogue, we are faced with a closure problem.

The simpler way to deal with the truncation problem is by ignoring correlations, assuming that the distribution of particles at (x_i,v_i) is not affected by particles at a distinct phase space point (x_j,v_j). In this mean field (or Hartree) approximation, the N-body Wigner function factorizes so that in a first approximation

$$f^{(2)}(x_1,v_1,x_2,v_2,t) = f(x_1,v_1,t)f(x_2,v_2,t). \tag{2.35}$$

Now (2.34) simplifies to

$$\frac{\partial f}{\partial t} + v_1 \frac{\partial f}{\partial x_1} = \int dv_1' \, K[W_{sc} \, | \, v_1' - v_1, x_1, t] \, f(x_1, v_1', t), \qquad (2.36)$$

with the mean field explicitly time-dependent self-consistent potential

$$W_{sc}(x,t) = \int dx' \, dv \, f(x', v, t) \, W(|x - x'|). \qquad (2.37)$$

The functional $K[W_{sc} \, | \, v_1' - v_1, x_1, t]$ is given by

$$K[W_{sc} \, | \, v_1' - v_1, x_1, t] = -\frac{im}{2\pi\hbar^2} \int ds_1 \, \exp\left(-\frac{im(v_1' - v_1)s_1}{\hbar}\right)$$
$$\times \left(W_{sc}\left(x_1 + \frac{s_1}{2}, t\right) - W_{sc}\left(x_1 - \frac{s_1}{2}, t\right)\right). \qquad (2.38)$$

In many cases, it is necessary to take into consideration an external, possibly time-dependent potential $V_{ext}(x_1, \ldots, x_N, t)$. For instance, such a circumstance arises in solid state devices, when considering the electronic motion in a fixed ionic lattice or under a confining field like in quantum wires or quantum wells [10, 20, 30]. Or even we can simply incorporate the field due to an homogeneous ionic background. In these cases, the external potential is of the form

$$V_{ext}(x_1, \ldots, x_N, t) = \sum_{i=1}^{N} W_{ext}(x_i, t) \qquad (2.39)$$

for some one-particle potential $W_{ext}(x_i, t)$. Implicitly in (2.39), the functional form of W_{ext} is the same irrespective of x_i, implying that the external field impose the same effect in all particles, which are supposed indistinguishable. Hence, for completeness, we indicate the changes for a potential

$$V(x_1, \ldots, x_N) = \sum_{i<j} W(|x_i - x_j|) + \sum_{i=1}^{N} W_{ext}(x_i, t). \qquad (2.40)$$

Repeating the steps in the derivation involving a self-consistent potential only, the time-evolution for the one-body reduced Wigner function $f(x_1, v_1, t)$ can then be shown to be governed by

$$\frac{\partial f}{\partial t} + v_1 \frac{\partial f}{\partial x_1} = \int dv_1' \, K[W_{sc} + W_{ext} \, | \, v_1' - v_1, x_1, t] \, f(x_1, v_1', t), \qquad (2.41)$$

where

$$K[W_{sc} + W_{ext} | v_1' - v_1, x_1, t] = -\frac{im}{2\pi\hbar^2} \int ds_1 \exp\left(-\frac{im(v_1' - v_1)s_1}{\hbar}\right)$$
$$\times \left(W_{sc}\left(x_1 + \frac{s_1}{2}, t\right) + W_{ext}\left(x_1 + \frac{s_1}{2}, t\right)\right.$$
$$\left. -W_{sc}\left(x_1 - \frac{s_1}{2}, t\right) - W_{ext}\left(x_1 - \frac{s_1}{2}, t\right)\right) \qquad (2.42)$$

and the averaged self-consistent potential W_{sc} is as in (2.37).

2.3 Electrostatic Quantum Plasmas

Consider now three-dimensional charged particle motion, with the Coulomb interaction

$$W(|\mathbf{r} - \mathbf{r}'|) = \frac{e^2}{4\pi\varepsilon_0 |\mathbf{r} - \mathbf{r}'|}, \qquad (2.43)$$

where $-e$ is the electron charge and ε_0 is the vacuum permittivity constant. In terms of the self-consistent W_{sc} and some external W_{ext} potentials, it is convenient to define the total electrostatic potential $\phi(\mathbf{r}, t)$ so that

$$\phi(\mathbf{r}, t) = \phi_{sc}(\mathbf{r}, t) + \phi_{ext}(\mathbf{r}, t), \qquad (2.44)$$

where

$$W_{sc}(\mathbf{r}, t) = -e\phi_{sc}(\mathbf{r}, t), \quad W_{ext}(\mathbf{r}, t) = -e\phi_{ext}(\mathbf{r}, t). \qquad (2.45)$$

It follows from the three-dimensional version of (2.37) that

$$\nabla^2 \phi_{sc} = -\frac{e}{\varepsilon_0} \int d\mathbf{r}' \, d\mathbf{v} \, f(\mathbf{r}', \mathbf{v}, t) \nabla^2 \left(\frac{1}{4\pi |\mathbf{r} - \mathbf{r}'|}\right)$$
$$= \frac{e}{\varepsilon_0} \int d\mathbf{r}' \, d\mathbf{v} \, f(\mathbf{r}', \mathbf{v}, t) \, \delta(\mathbf{r} - \mathbf{r}')$$
$$= \frac{e}{\varepsilon_0} \int d\mathbf{v} \, f(\mathbf{r}, \mathbf{v}, t). \qquad (2.46)$$

Moreover,

$$\nabla^2 \phi_{ext} = -\frac{1}{e} \nabla^2 W_{ext} \equiv -\frac{n_0 e}{\varepsilon_0} \qquad (2.47)$$

if the external potential is caused by an immobile fixed homogeneous ionic background of density n_0 and ion charge e. Appropriate changes are needed in the case of a nonhomogeneous background, for example, as in the case of doped semiconductors, or in presence of a dispersive medium with a permittivity constant $\varepsilon \neq \varepsilon_0$ [10, 20, 30].

Combining (2.46) and (2.47), it is immediate to obtain

$$\nabla^2 \phi = \frac{e}{\varepsilon_0} \left(\int d\mathbf{v} \, f(\mathbf{r}, \mathbf{v}, t) - n_0 \right),$$
(2.48)

which is the Poisson equation in this case.

Just for notational simplicity, it is better to restrict again to the one-dimensional case. In this way the expressions look nicer, and the transition to three spatial dimensions can be easily done if necessary. In terms of the electrostatic potential ϕ, (2.41) is rephrased as

$$\frac{\partial f}{\partial t} + v \frac{\partial f}{\partial x} = \int dv' \, K_\phi[\phi \,|\, v' - v, x, t] \, f(x, v', t),$$
(2.49)

where $K_\phi[\phi \,|\, v' - v, x, t]$ is the following functional,

$$K_\phi[\phi \,|\, v' - v, x, t] = \frac{iem}{\hbar} \int \frac{ds}{2\pi\hbar} \exp\left(\frac{im(v' - v)s}{\hbar} \right)$$
$$\times \left(\phi\left(x + \frac{s}{2}, t\right) - \phi\left(x - \frac{s}{2}, t\right) \right).$$
(2.50)

Equation (2.49) can be termed the quantum Vlasov equation (in the electrostatic case), since it is the quantum analog of the Vlasov equation satisfied by the reduced one-particle distribution function. The quantum Vlasov equation for the Wigner function should be coupled to the Poisson equation for the scalar potential,

$$\frac{\partial^2 \phi}{\partial x^2} = \frac{e}{\varepsilon_0} \left(\int dv \, f(x, v, t) - n_0 \right).$$
(2.51)

Equations (2.49) and (2.51) constitute the Wigner–Poisson system, which is the fundamental model for electrostatic quantum plasmas. It determines in a self-consistent way both the Wigner function, associated with how the particles distribute in phase space, and the scalar potential, which in turn describe the forces acting on the particles.

Equations (2.49) and (2.51) need to be supplemented with suitable boundary and initial conditions. For plasmas, frequently decaying or periodic boundary conditions are sufficient. For nano-devices, the choice of boundary conditions is subtler due to the finite size of the system and the nonlocal character of the Wigner function. Indeed, to compute the integral defining the Wigner function, we need to specify $f(x, v, 0)$ in the whole space even when dealing with finite size systems. We refer to the specialized literature for more details [10, 20, 30].

Before seeking some of the consequences of the Wigner–Poisson system, let us recapitulate the steps toward its derivation. First, it is a mean field model with the N-body ensemble Wigner function supposed to be factorisable. Thanks to this property, we achieved the simplest solution to the closure problem of the quantum

BBGKY hierarchy. In particular, it follows a notable advantage over the N-body Schrödinger equation (or equivalently over the Liouville–von Neumann equation for the N-body ensemble density matrix): the tremendous reduction of the number of independent variables. For $N \gg 1$ electrons in three-dimensional space, we can compare the $3N + 1$ coordinates entering the wavefunction, the $6N + 1$ coordinates of the density matrix, and the $6 + 1 = 7$ independent variables of the reduced one-body Wigner function, taking into account time. In particular, the mean field theory is much less numerically demanding, since it requires the discretization of a space with fewer dimensions. However, since correlations are disregarded, the Wigner–Poisson system does not incorporate collisions. Moreover, no spin or relativistic effects are taken into account in our presentation. Finally, no magnetic fields were introduced yet.

Once the Wigner–Poisson system has been derived, it becomes the natural tool in quantum kinetic theory for plasmas, since it is exactly analog to the Vlasov–Poisson system. Hence, the methods applied to the Vlasov–Poisson system can with some optimism be directly translated to quantum plasmas. Other quantum kinetic treatments for assemblies of charged particle systems are obviously important, but cannot compete with the Wigner formalism in the quantum plasma context. For instance, the density functional [10] and Green's function [17, 21] approaches are popular tools for the modeling of quantum transport in the condensed matter community. However, presently the majority of the plasma physics researchers feels more comfortable with the particle distribution function method where the Vlasov equation plays a central rôle and for which a number of analytical and numerical methods are already available. Nevertheless, the simplifications underlying the Wigner–Poisson model points to the relevance of the alternative approaches toward a more sophisticated modeling. For instance, using Green's function techniques to describe collisions associated with short range particle–particle interactions one finds [17,21] a nonlocal Boltzmann type collision operator which has to be included in (2.49). However, these developments are outside the scope of the present text.

It is instructive to analyze the semiclassical limit of the quantum Vlasov equation (2.49). By means of the change of variable $s = \hbar \tau / m$ and Taylor expanding, the result is

$$\frac{\partial f}{\partial t} + v\frac{\partial f}{\partial x} - \frac{eE}{m}\frac{\partial f}{\partial v} = -\frac{e\hbar^2}{24\,m^3}\frac{\partial^2 E}{\partial x^2}\frac{\partial^3 f}{\partial v^3} + O(H^4), \qquad (2.52)$$

where the electric field is $E = E(x,t) = -\partial\phi/\partial x$. Implicitly, the semiclassical approximation assumes the smallness of a nondimensional quantum parameter $H = \hbar/(mv_0 L_0)$, where v_0 and L_0 are, respectively, typical velocity and length scales.

If no quantum effects were present, (2.52) reduces to Vlasov's equation,

$$\frac{\partial f}{\partial t} + v\frac{\partial f}{\partial x} - \frac{eE}{m}\frac{\partial f}{\partial v} = 0, \qquad (2.53)$$

to be coupled to Poisson's equation to compose the Vlasov–Poisson system.

As an intermediate step in the derivation of (2.52), one need to calculate integrals like

$$\int \frac{\mathrm{d}v'\,\mathrm{d}\tau}{2\pi}\,\tau\,\mathrm{e}^{\mathrm{i}(v'-v)\tau} f(x,v',t) = \mathrm{i}\frac{\partial}{\partial v}\int \frac{\mathrm{d}v'\,\mathrm{d}\tau}{2\pi}\,\mathrm{e}^{\mathrm{i}(v'-v)\tau} f(x,v',t)$$

$$= \mathrm{i}\frac{\partial}{\partial v}\int \mathrm{d}v'\,\delta(v'-v)\,f(x,v',t) = \mathrm{i}\frac{\partial f}{\partial v}. \qquad (2.54)$$

It is immediate to recognize (2.52) as a semiclassical Vlasov equation, with f playing the rôle of one-particle distribution function. From the Wigner function, one can compute macroscopic quantities like particle, current and energy densities, very much like in classical physics. Hence, it is a natural trend, to investigate to which extent the methods applied to the Vlasov–Poisson system can be useful in the Wigner–Poisson context. However, unlike in the classical limit, in general neither f nor phase space volume are preserved by the quantum Vlasov equation, since

$$\frac{\mathrm{d}f}{\mathrm{d}t} = -\frac{e\hbar^2}{24m^3}\frac{\partial^2 E}{\partial x^2}\frac{\partial^3 f}{\partial v^3} + O(H^4) \neq 0 \qquad (2.55)$$

along the (classical) characteristic equations $\dot{x} = v, \dot{v} = -eE/m$. Moreover the positive definiteness of the Wigner function is not preserved by (2.49). The exception is for linear electric fields, for which the quantum corrections vanishes in (2.52). In this case, the Wigner and Vlasov equations coincide to all orders in the nondimensional quantum parameter H.

Even in the harmonic oscillator case, when the Wigner and Vlasov equations coincide, $f(x,v,t)$ cannot be considered as an ordinary probability distribution function. Indeed, not all functions on phase space can be taken as Wigner functions, since a genuine Wigner function should correspond to a positive definite density matrix. Therefore, at least the following necessary conditions [18] must hold,

$$\int \mathrm{d}x\mathrm{d}v f = N, \qquad (2.56)$$

$$\int \mathrm{d}v f \geq 0, \qquad (2.57)$$

$$\int \mathrm{d}x f \geq 0, \qquad (2.58)$$

$$\int \mathrm{d}x\mathrm{d}v f^2 \leq \frac{mN^2}{2\pi\hbar}. \qquad (2.59)$$

Equation (2.56) is just a normalization condition, while (2.57) and (2.58) arise because the spatial and velocity marginal probability densities should be everywhere nonnegative. Finally, (2.59) is needed to avoid violation of the uncertainty principle, eliminating too spiky functions $f(x,v,t)$.

2.4   The Schrödinger–Poisson System

The equivalence between the Wigner–Poisson and a system of countably many Schrödinger equations coupled to the Poisson equation has been mathematically demonstrated [29]. More exactly, any Wigner function can be written as

$$f(x,v,t) = \frac{Nm}{2\pi\hbar} \sum_{\alpha=1}^{M} p_\alpha \int ds \exp\left(\frac{imvs}{\hbar}\right) \psi_\alpha^*\left(x+\frac{s}{2},t\right) \psi_\alpha\left(x-\frac{s}{2},t\right), \quad (2.60)$$

with ensemble probabilities $p_\alpha \geq 0$ so that $\sum_{\alpha=1}^{M} p_\alpha = 1$, for each one-particle ensemble wavefunctions $\psi_\alpha(x,t)$ satisfying

$$i\hbar\frac{\partial \psi_\alpha}{\partial t} = -\frac{\hbar^2}{2m}\frac{\partial^2 \psi_\alpha}{\partial x^2} - e\phi\psi_\alpha, \quad \alpha = 1,\ldots,M, \quad (2.61)$$

which is the Schrödinger equation for a particle under the action of the mean field potential $\phi(x,t)$. In addition, the Poisson equation (2.51) is rewritten as

$$\frac{\partial^2 \phi}{\partial x^2} = \frac{e}{\varepsilon_0}\left(N\sum_{\alpha=1}^{M} p_\alpha |\psi_\alpha(x,t)|^2 - n_0\right). \quad (2.62)$$

Equations (2.61) and (2.62) constitute the so-called Schrödinger–Poisson system, which has to be supplemented with suitable initial and boundary conditions. It provides a way of replacing the original N-body problem by a collection of one-body problems, coupled by Poisson's equation. From a methodological point of view, the Schrödinger–Poisson modeling corresponds to put the emphasis again on the wavefunction and not on the (phase space) Wigner function. Collective effects are mediated by the self-consistent potential ϕ.

A rigorous proof [29] of the equivalence of (2.61) and (2.62) and the Wigner–Poisson system (2.49)–(2.51) is beyond the present text. However, we can obtain some insight on the interpretation of the ensemble wavefunctions. From (2.12)–(2.20),

$$f(x_1,v_1,t) = \int dx_2 dv_2,\ldots,dx_N dv_N\, f^N(x_1,v_1,\ldots,x_N,v_N,t)$$

$$= N\left(\frac{m}{2\pi\hbar}\right)\sum_{\alpha=1}^{M} p_\alpha \int ds_1\, dx_2,\ldots,dx_N \exp\left(\frac{imv_1 s_1}{\hbar}\right)$$

$$\times \psi_\alpha^{N*}\left(x_1+\frac{s_1}{2},x_2,\ldots,x_N,t\right)\psi_\alpha^{N}\left(x_1-\frac{s_1}{2},x_2,\ldots,x_N,t\right). \quad (2.63)$$

To follow the mean field approximation, we are tempted to assume the factorized form

$$\psi_\alpha^{N}(x_1,x_2,\ldots,x_N,t) = \psi_\alpha(x_1,t) \times \cdots \times \psi_\alpha(x_N,t), \quad (2.64)$$

for the N-body wavefunction, fully neglecting the correlations due to the interaction potential. Quantum statistics effects are not taken into account in the *Ansatz* (2.64), which does not respect the Pauli principle. However, for simplicity spin considerations will be not included at this moment. Inserting (2.64) into (2.63), the result is precisely (2.60), with the same statistical weights p_α. Hence we can view the one-body ensemble wavefunctions $\psi_\alpha(x,t)$ as the result of splitting the N-body ensemble wavefunction into the product of identical factors.

Equation (2.60) shows that the reduced one-body Wigner function is always in the form of the sum

$$f(x,v,t) = \sum_{\alpha=1}^{M} p_\alpha f_\alpha(x,v,t), \tag{2.65}$$

where

$$f_\alpha(x,v,t) = \frac{Nm}{2\pi\hbar} \int ds \, \exp\left(\frac{imvs}{\hbar}\right) \psi_\alpha^*\left(x+\frac{s}{2},t\right) \psi_\alpha\left(x-\frac{s}{2},t\right). \tag{2.66}$$

The case where only one ensemble wavefunction is needed so that $p_\alpha = \delta_{\alpha\beta}$ for some β corresponds to a pure state. Otherwise, we have a mixed state.

Using the map (2.7) from the Wigner function to the density matrix $\rho(x,y,t)$, we can derive a condition for a pure state. Supposing

$$f(x,v,t) = \frac{Nm}{2\pi\hbar} \int ds \, e^{\frac{imvs}{\hbar}} \psi^*\left(x+\frac{s}{2},t\right) \psi\left(x-\frac{s}{2},t\right) \tag{2.67}$$

in terms of a single wavefunction $\psi(x,t)$ and inserting in (2.7), we obtain

$$\rho(x,y,t) = N\psi(x,t)\psi^*(y,t). \tag{2.68}$$

Hence,

$$\frac{\partial^2 \ln\rho(x,y,t)}{\partial x \partial y} = 0 \tag{2.69}$$

is a necessary condition for a pure state, where $\rho(x,y,t)$ is given by (2.7). In addition to (2.69), a real Wigner function is required to qualify a pure state, which means $\rho^*(x,y,t) = \rho(y,x,t)$. Moreover, it can be shown that a pure state at $t = 0$ remains a pure state along the time-evolution of the quantum Vlasov equation (2.49).

While the direct construction using (2.60) of the Wigner function from the wavefunctions and the statistical weights is a trivial task, the reverse problem of how to choose ψ_α, p_α to reproduce a given Wigner function is more obscure. In particular, when we know we are dealing with a mixed state, what is the minimal number M of ensemble wavefunctions needed? Or, for a fixed M, what could be the natural way to define the wavefunctions ψ_α and the corresponding probabilities p_α so as to reproduce $f(x,v,t)$, possibly in some approximate sense? There is no universal answer to these questions in the current literature.

2.5 Validity of the Wigner–Poisson System

The Wigner–Poisson system is collisionless, in the same sense as is the Vlasov–Poisson system of classical plasma physics. In both models, the long-range interactions due to the self-consistent electrostatic potential are assumed to dominate over short-range collisional interactions between two or more particles. This statement can be made more precise [31]. Correlations between particles or equivalently collisions cannot be neglected if the average potential energy between two electrons become comparable to the average kinetic energy. As have been seen in the discussion on the classical and quantum energy coupling parameters of Sect. 1.4, we know the validity conditions for the collisionless approximation. For classical plasmas, it reads

$$\Gamma_C = \frac{U_{\text{pot}}}{K_C} = \frac{e^2 n_0^{1/3}}{\varepsilon_0 \kappa_B T} \ll 1, \tag{2.70}$$

in terms of the classical coupling parameter Γ_C. On the other hand, for plasmas where the Fermi–Dirac statistics is unavoidable, collisions can be ignored provided

$$\Gamma_Q = \frac{U_{\text{pot}}}{K_Q} = \frac{e^2 n_0^{1/3}}{\varepsilon_0 \kappa_B T_F} \sim \frac{m e^2}{\varepsilon_0 \hbar^2 n_0^{1/3}} \ll 1, \tag{2.71}$$

in terms of the quantum coupling parameter Γ_Q. As a consequence, for quantum plasmas ($T_F > T$) the collisionless approximation becomes better as the density increases.

When the condition $\Gamma_C \ll 1$ holds, the N-particle distribution can be factorized as a product of one-particle distribution functions satisfying Vlasov's equation. Hence the Vlasov–Poisson system is the standard model to describe classical electrostatic plasmas in the collisionless approximation.

When the condition $\Gamma_Q \ll 1$ holds, a quantum electron gas can be described by the Wigner–Poisson system. In this case, the N-body Wigner function is expressed as a product of one-particle Wigner functions so that the Wigner–Poisson system is the natural model for collisionless quantum plasmas.

The previous results were derived in the limiting cases $T \gg T_F$ (classical) and $T \ll T_F$ (quantum degenerate). For intermediate temperatures, simple expressions for the coupling parameters are not available, but one must expect a smooth transition between the two regimes.

For electrons in metal, we have typically

$$n_0 \simeq 10^{29}\,\text{m}^{-3}, \quad v_F \simeq 10^6\,\text{ms}^{-1}, \quad \omega_p \simeq 10^{16}\,\text{s}^{-1}, \quad \lambda_F \simeq 10^{-10}\,\text{m}. \tag{2.72}$$

These values yield a quantum coupling parameter of order unity. Allowing for the dimensionless constants, we have neglected and the different properties of metals, we realize that Γ_Q can be both smaller and larger than unity for typical metallic electrons [31].

Since $\Gamma_Q \sim 1$, apparently a collisionless model such as the Wigner–Poisson system could not be employed for metals. However, fortunately, the average rate of electron–electron collisions in such system is drastically reduced due to the Fermi–Dirac statistics. Indeed, in most cases of interest, for relatively low temperatures the vast majority of electrons is well below the Fermi energy. Since all lower levels are occupied, the exclusion principle forbids transitions except for the small electron population in a shell of thickness $\sim \kappa_B T$ around the Fermi surface, a phenomena know as Pauli blocking [2]. The e–e collision rate (inverse of the lifetime τ_{ee}) for such electrons is proportional to $\kappa_B T / \hbar$, as a consequence of the uncertainty principle energy $\times$ time $\sim \hbar$. Since the fraction of electrons available to collisions is $\sim T / T_F$, one obtains

$$\frac{1}{\tau_{ee}} \sim \frac{1}{\hbar} \frac{\kappa_B T^2}{T_F}. \tag{2.73}$$

At room temperature, $\tau_{ee} \simeq 10^{-10}\,\mathrm{s}$, which is much larger than the typical collisionless time scale $\tau_p = \omega_p^{-1} \simeq 10^{-16}\,\mathrm{s}$. Therefore, for times smaller than τ_{ee}, the effect of e–e collisions can be safely neglected. In addition, it turns out that the typical relaxation time scale is $\tau_r \simeq 10^{-14}\,\mathrm{s}$, which is again significantly larger than τ_p. In summary, the ordering

$$\tau_p \ll \tau_r \ll \tau_{ee}, \tag{2.74}$$

implies that a collisionless (Wigner) model is appropriate for relatively short time scales [31].

Notice that not only very dense charged particle systems deserve quantum kinetic equations for their description. For instance, due to the ongoing miniaturization, even scarcely populated electronic systems such as resonant tunneling diodes [30] should be described in terms of quantum models. Indeed, the behavior of these ultra-small electronic devices relies on quantum diffraction effects as tunneling, making purely classical methods inappropriate. The nonlocal integro-differential potential term in (2.49) in the Wigner–Poisson system has been shown to be capable of the modeling of negative differential resistance, associated with tunneling [25]. Moreover, the collisionless approximation become more reasonable in view of the nanometric scale of the devices, simply because the mean free-path exceeds the system size. In the same manner, the usually extreme high operating frequencies makes the collisionless approximation more accurate because $\omega \tau \ll 1$ for an operating frequency ω and a average time τ between collisions. For example, in resonant tunneling diodes one can find [30] potential barriers of the order $0.3\,\mathrm{eV} \sim \hbar \omega$, implying an operating frequency $\omega \sim 10^{15}\,\mathrm{s}^{-1}$. Therefore, the Wigner–Poisson system is well suited for ballistic, collisionless processes in nanometric solid state devices, even at relatively low densities of order $n_0 \sim 10^{24}\,\mathrm{m}^{-3}$. Correspondingly, one finds a Fermi temperature $T_F \sim 40\,\mathrm{K}$ much smaller than a typical room temperature $T \sim 300\,\mathrm{K}$, justifying the nondegeneracy assumption and Maxwell–Boltzmann's statistics.

2.6 Extensions to Include Correlation and Spin Effects

In spite of $df/dt \neq 0$, neither (2.49) nor (2.52) include collisions. Instead, the mean field (or Hartree) approximation is implicit in the Wigner–Poisson model, since its derivation assumed the factorization of the N-particle Wigner function $f^N = f^N(x_1, v_1, \ldots, x_N, v_N, t)$ as a product of N identical one-particle Wigner functions, $f^N = f(x_1, v_1, t) \times \cdots \times f(x_N, v_N, t)$. In this context, the scalar potential $\phi(x, t)$ comes from the collective field of the N electrons, just as in the classical theory. Allowing for correlations would result in a quantum BBGKY hierarchy.

In principle, a more detailed factorization taking into account the Pauli exclusion principle could have been employed. In this case, a Hartree–Fock term would be present in (2.49), see [24] for more details. However, frequently it is expedite to replace the complicated, nonlocal exchange-correlation terms by local phenomenological expressions, using the so-called adiabatic local density approximation (ALDA) [13, 33].

As remarked, quantum effects in a plasma (and in N-body systems in general) are unavoidable when the particle density is high enough. This can also be seen through the expansion parameter H in (2.52), which increase with density. To verify this, assume L_0 of the order of the mean inter-particle distance $n_0^{-1/3}$ and v_0 of the order of the thermal velocity v_T. Hence, $H \sim \hbar n_0^{1/3}/(mv_T)$, which is of order unity when the de Broglie wavelength $\lambda_B = \hbar/(mv_T)$ is comparable to the mean inter-particle separation. In this case, there will be a significant overlap of the wave packets associated with each electron, so that the Newtonian approximation breaks down. On the other hand, a collisionless model for quantum plasmas becomes more accurate for higher densities, see the quantum coupling parameter Γ_Q in (2.71). We use this conclusion as a methodological argument in favor of the Wigner–Poisson system, even if the underlying Fermi–Dirac statistics is not included in (2.49) and (2.51). We also note that the classical energy coupling parameter usually Γ_C in (2.70) plays a marginal rôle in quantum plasmas.

On the other hand, it is worth to say that the Wigner–Poisson system is employed in the semiconductor literature where typically the particle densities are not so high. Except for short-time ballistic phenomena, in such cases, it is crucial to improve the model by means of adequate collision operators [30]. For instance, in a resonant tunneling diode quantum effects are noticeable thanks to the smallness of the system, so that the basic characteristic length is the size of the device rather than the average inter-particle distance. Mathematically, the size of the system manifests, for example, through the boundary conditions.

In recent years [1, 9, 28], much attention has been devoted to the Wigner–Fokker–Planck equation

$$\frac{\partial f}{\partial t} + v \frac{\partial f}{\partial x} - \int dv' \, K_\phi[\phi \,|\, v' - v, x, t] f(v', x, t) = L_{\text{QFP}}[f], \tag{2.75}$$

for the modeling of quantum dissipation, with

$$L_{\text{QFP}}[f] = \frac{D_{pp}}{m^2} \frac{\partial^2 f}{\partial v^2} + \frac{\eta}{m} \frac{\partial}{\partial v}(vf) + \frac{2}{m} D_{pq} \frac{\partial^2 f}{\partial v \partial x} + D_{qq} \frac{\partial^2 f}{\partial x^2} \tag{2.76}$$

acting as a collision term. In this case, it is assumed an open quantum system interacting with a heat bath of harmonic oscillators. Here,

$$D_{pp} = \eta \, \kappa_B T, \quad D_{pq} = \frac{\eta \, \Omega \, \hbar^2}{12 \, \pi \, m \kappa_B T}, \quad D_{qq} = \frac{\eta \, \hbar^2}{12 \, m^2 \, \kappa_B T} \tag{2.77}$$

are phenomenological constants related to the interactions, where η is the damping coefficient of the bath, T is the bath temperature and Ω is the cut-off frequency of the reservoir oscillators. Physically, the heat bath of harmonic oscillators can be realized, for example, in terms of the phonons propagating in a crystal lattice.

Notice that for homogeneous equilibrium ($\partial f / \partial x = 0$) or for vanishing quantum effects, the Maxwellian $f = f_M \sim \exp(-v^2/2v_T^2), v_T^2 = \kappa_B T/m$ belongs to the kernel of the collision operator (2.76), or

$$L_{QFP}[f_M] = 0. \tag{2.78}$$

Actually this comes with no surprise since the Wigner–Fokker–Planck model is derived [1, 9] on the assumption of classical statistics, that is, no spin degrees of freedom. In addition, if D_{pq} and D_{qq} are set to zero the Caldeira–Legget [7] model is recovered. However, importantly the Wigner–Fokker–Planck collision operator can be put in the Lindblad form [27] provided

$$D_{pp} D_{qq} \geq D_{pq}^2 + \frac{\hbar^2 \eta^2}{16 m^2} \quad \text{or} \quad \hbar \Omega \leq \sqrt{3} \pi \kappa_B T, \tag{2.79}$$

the last inequality holding for $\eta \neq 0$. Accordingly, it can be shown that the associated density matrix operator preserves positivity under time-evolution, a feature not satisfied by the Caldeira–Legget model.

Promising as it is, the Wigner–Fokker–Planck model did not apply to dense quantum plasma astrophysical environments, where the fermion statistics play a significant rôle. Therefore, the inclusion of suitable dissipation mechanisms is a challenge in quantum plasma physics.

2.7 High Frequency Longitudinal Waves

For any plasma, the propagation of linear waves is an essential issue. Assuming a wave with wave number k and frequency ω propagating in a plasma described by the Wigner–Poisson system (2.49) and (2.51), set

$$f(x, v, t) = f_0(v) + f_1(v) \exp(i[kx - \omega t]), \tag{2.80}$$

$$\phi = \phi_1 \exp(i[kx - \omega t]), \tag{2.81}$$

for first-order disturbances f_1, ϕ_1. It is supposed an equilibrium Wigner function $f = f_0(v)$ such that $\int dv \, f_0(v) = n_0$ and a zero equilibrium electrostatic potential.

Linearizing (2.49) and (2.51) it follows

$$-i(\omega - kv)f_1(v) = \left(f_0\left(v + \frac{\hbar k}{2m}\right) - f_0\left(v - \frac{\hbar k}{2m}\right) \right)\phi_1, \qquad (2.82)$$

$$-k^2\phi_1 = \frac{e}{\varepsilon_0}\int dv f_1(v). \qquad (2.83)$$

This linear homogeneous system for f_1, ϕ_1 admit nontrivial solutions if and only if

$$\varepsilon \equiv 1 - \frac{m\,\omega_p^2}{n_0\hbar k^2}\int dv\,\frac{f_0[v + \hbar k/(2m)] - f_0[v - \hbar k/(2m)]}{kv - \omega} = 0. \qquad (2.84)$$

Here $\omega_p = (n_0 e^2/m\varepsilon_0)^{1/2}$ is the plasma frequency.

For the dispersion properties only, (2.84) can be understood in the principal value sense. In addition, it is convenient to change integration variables so as to rewrite the permittivity ε as

$$\varepsilon = 1 - \frac{\omega_p^2}{n_0}\int dv\,\frac{f_0(v)}{(\omega - kv)^2 - \hbar^2 k^4/(4m^2)}. \qquad (2.85)$$

Further, it is useful [14] to introduce the rescaling

$$F = \frac{\omega_p f_0}{n_0 k}, \quad u = \frac{kv}{\omega_p}, \quad \Omega = \frac{\omega}{\omega_p}, \qquad (2.86)$$

so that

$$\varepsilon = 1 - \frac{1}{\Omega^2}\int \frac{du\,F(u)}{(1 - u/\Omega)^2 - \Omega_q^2/\Omega^2} = 0, \qquad (2.87)$$

where $\int du\,F(u) = 1$ and it was defined

$$\Omega_q^2 = \frac{\hbar^2 k^4}{4m^2\,\omega_p^2}. \qquad (2.88)$$

For high frequency oscillations, we can consider expanding the integrand retaining up to $O\left(\Omega_q^2/\Omega^2, \langle u^2\rangle/\Omega^2\right)$ terms, (2.87) becomes

$$\varepsilon \simeq 1 - \frac{1}{\Omega^2}\left(1 + \frac{3\,\langle u^2\rangle + \Omega_q^2}{\Omega^2}\right) = 0. \qquad (2.89)$$

In (2.89), $\langle u^2\rangle = \int du\,F(u)\,u^2$. For simplicity, $\int du\,F(u)\,u = 0$ was assumed, which holds for instance for symmetric equilibria.

Solving (2.89) by successive approximations, the result is

$$\Omega^2 = 1 + 3 \langle u^2 \rangle + \Omega_q^2 \tag{2.90}$$

or

$$\omega^2 = \omega_p^2 + 3k^2 \langle v^2 \rangle + \frac{\hbar^2 k^4}{4m^2}, \tag{2.91}$$

where $\langle v^2 \rangle = \int dv\, f_0(v)\, v^2/n_0$. Equation (2.91) is the Bohm–Pines dispersion relation [5], the quantum counterpart of the Bohm–Gross dispersion relation of classical high frequency longitudinal plasma waves [4].

Equation (2.91) describes quantum Langmuir waves and is correct no matter the form of the equilibrium Wigner function, as far as the high frequency hypothesis is valid. In the case [24] of zero velocity dispersion ($f_0(v) = n_0\, \delta(v)$) one has

$$\omega^2 = \omega_p^2 + \frac{\hbar^2 k^4}{4m^2}. \tag{2.92}$$

Here, it is referred to "zero velocity dispersion" rather than to "zero-temperature" to not confound with, for instance, a zero-temperature Fermi gas, where $\langle v^2 \rangle \neq 0$ in consequence of the exclusion principle.

For the imaginary part of the frequency, it is useful to rewrite (2.84) according to

$$\varepsilon = 1 - \frac{m\,\omega_p^2}{n_0 \hbar k^2} \left(\int_{L_+} \frac{dv\, f_0(v)}{k[v - \hbar k/(2m)] - \omega} - \int_{L_-} \frac{dv\, f_0(v)}{k[v + \hbar k/(2m)] - \omega} \right) = 0, \tag{2.93}$$

where the velocity integrals are performed with Landau contours $L_\pm$ passing under the poles at $v = \omega/k \pm \hbar k/(2m)$. Equation (2.93) can be used [24, 35] as the starting point for the discussion of the quantum Landau damping, the quantum counterpart of the collisionless damping present in classical plasmas.

From (2.93), assuming that the damping or growth rate γ is small and repeating the procedure for classical plasmas (see Problem 2.7), we get

$$\gamma = \frac{\pi\, \omega_p^3}{4 n_0 k^2} \left(\frac{f_0[\omega/k + \hbar k/(2m)] - f_0[\omega/k - \hbar k/(2m)]}{\hbar k/(2m)} \right), \tag{2.94}$$

where ω follows from (2.90). Hence, the damping (or growth) rate of quantum Langmuir waves is a finite-difference version of the classical growth rate γ_{cl}, which is obtained from (2.94) in the formal limit $\hbar \to 0$,

$$\gamma_{cl} = \frac{\pi\, \omega_p^3}{2 n_0 k^2} \frac{df_0}{dv} \left(v = \frac{\omega}{k} \right). \tag{2.95}$$

It is relevant to remark that from (2.94), a particular class of stationary solutions such that $\gamma = 0$ is given by any function $f_0(v)$ which is a periodic in velocity space, with period $\hbar k/m$:

$$f_0\left(\frac{v+\hbar k}{2m}\right) - f_0\left(\frac{v-\hbar k}{2m}\right) = 0. \tag{2.96}$$

Assuming $f_0(v) \sim \exp(i\alpha v)$ in (2.96), with α to be determined, we obtain the characteristic equation $\sin(\alpha \hbar k/(2m)) = 0$. Hence, an exact equilibrium solution is the linear combination, or Fourier series

$$f_0(v) = a_0 + \sum_{n=1}^{\infty} a_n \cos\left(\frac{2\pi n v}{\lambda_v}\right) + \sum_{n=1}^{\infty} b_n \sin\left(\frac{2\pi n v}{\lambda_v}\right), \tag{2.97}$$

where a_n, b_n are arbitrary real constants and $\lambda_v = \hbar k/m$. Notice the singular character of the quantum oscillations, whose "wavelength" of the fundamental mode ($n = 1$) in velocity space tends to zero as $\hbar \to 0$. The solution given by (2.97) represents periodic oscillations in velocity space. This is in sharp contrast to the classical stationary solution which points to the formation of a plateau ($df_0/dv = 0$) at the resonance, see (2.95). The oscillating character of quantum plasma equilibria has been predicted in terms of a quantum quasilinear theory and numerically verified [16].

The treatment of nonlinear phenomena in the Wigner–Poisson framework is rather involved, with few known analytic results. For instance, a few classes of exact nonlinear stationary solution is available [15]. The most expedite route toward nonlinear quantum plasmas is by means of hydrodynamic formulations, as we start to verify in the next chapter.

Problems

2.1. Use $v \exp(-imvs/\hbar) = (i\hbar/m)(\partial/\partial_s) \exp(-imvs/\hbar)$ and integration by parts assuming decaying or periodic boundary conditions to check (2.3).

2.2. Check the last equality in (2.7) as well as (2.8).

2.3. Derive (2.52) starting from (2.49).

2.4. Work out (2.56) and (2.59) for a Gaussian shaped Wigner function $f = A \exp(-v^2/\delta v^2 - x^2/\delta x^2), A = $ cte. Show that $m\,\delta v\,\delta x \geq \hbar$.

2.5. Expand the equation of motion for the one-particle Wigner function in one spatial dimension up to fourth-order in the dimensionless quantum parameter H. Discuss the properties of quantum robust solutions, defined as the Wigner functions for which the quantum effects vanish up to $O(H^5)$.

2.6. Show that the Maxwellian belongs to the kernel of the Wigner–Fokker–Planck collision operator.

2.7. Consult a standard plasma textbook where the damping rate of Langmuir waves is derived and repeat the procedure in the quantum case, to get (2.94) using (2.93).

References

1. Arnold, A., López, J. L., Markowich, P., Soler, J.: An analysis of quantum Fokker-Planck models: a Wigner function approach. Rev. Mat. Iberoamericana **20**, 771–814 (2004)
2. Ashcroft, N. W. and Mermin, N. D.: Solid state physics. Saunders College Publishing, Orlando (1976)
3. Bogoliubov, N. N.: Kinetic equations. J. Exp. Theor. Phys. **16**, 691–702 (1946)
4. Bohm, D. and Gross, E.: Theory of plasma oscillations. A. Origin of medium-like behavior. Phys. Rev. **75**, 1851–1864 (1949)
5. Bohm, D., Pines, D.: A collective description of electron interactions: III. Coulomb interactions in a degenerate electron gas. Phys. Rev. **92**, 609–625 (1953)
6. Born, M., Green, H. S.: A general kinetic theory of liquids I. The molecular distribution functions. Proc. Roy. Soc. A **188**, 10–18 (1946)
7. Caldeira, A. O., Leggett, A. J.: Path integral approach to quantum Brownian motion. Physica A **121**, 587–616 (1983)
8. Carruthers, P., Zachariasen, F.: Quantum collision theory with phase-space distributions. Rev. Mod. Phys. **55**, 245–285 (1983)
9. Castella, F., Erdõs, L., Frommlet, F., Markowich, P.: Fokker-Planck equations as scaling limits of reversible quantum systems. J. Stat. Phys. **100**, 543–601 (2000)
10. DiVentra, M.: Electrical Transport in Nanoscale Systems. Cambridge, New York (2008)
11. Glauber, R. J.: Coherent and incoherent states of the radiation field. Phys. Rev. **131**, 2766–2788 (1963)
12. Glauber, R.J.: Optical coherence and photon statistics. In: Dewitt, C., Blandin, A., Cohen-Tannoudji, C. (eds.) Quantum Optics and Electronics, pp. 63–185, Gordon and Breach, New York (1965)
13. Gusev, G. M., Quivy, A. A., Laman, T. E., Leite, J. R., Bakarov, A. K., Topov, A. I., Estibals, O., Portal, J. C.: Magnetotransport of a quasi-three-dimensional electron gas in the lowest Landau level. Phys. Rev. B **65**, 205316–205325 (2002)
14. Haas, F.: On quantum plasma kinetic equations with a Bohmian force. J. Plasma Phys. **76**, 389–393 (2010)
15. Haas, F. and Shukla, P. K.: Nonlinear stationary solutions of the Wigner and Wigner–Poisson equations. Phys. Plasmas **15**, 112302-112302-6 (2008)
16. Haas, F., Eliasson, B., Shukla, P. K. and Manfredi, G.: Phase-space structures in quantum-plasma wave turbulence. Phys. Rev. E **78**, 056407–056414 (2008)
17. Haug, H. J. W., Jauho, A. P.: Quantum Kinetics in Transport and Optics of Semiconductors. Springer, Berlin-Heidelberg (2008)
18. Hillery, M., O'Connell, R. F., Scully, M. O., Wigner, E. P.: Distribution functions in physics - fundamentals. Phys. Rep. **106**, 121–330 (1990)
19. Husimi, K.: Some formal properties of the density matrix. Prog. Phys. Math. Soc. Japan **22**, 264–314 (1940)
20. Jungel, A.: Transport Equations for Semiconductors. Springer, Berlin-Heidelberg (2009)
21. Kadanoff, L.P., Baym, G.: Quantum Statistical Mechanics: Green's Function Methods in Equilibrium and Non-Equilibrium Problems. Benjamin, New York (1962)
22. Kirkwood, J. G.: Quantum statistics of almost classical assemblies. Phys. Rev. **44**, 31–37 (1933)
23. Kirkwood, J. G.: The statistical mechanical theory of transport processes I. General theory. J. Chem. Phys. **14**, 180–201 (1946)
24. Klimontovich Y. and Silin, V. P.: The spectra of systems of interacting particles. In: Drummond, J. E. (ed.) Plasma Physics, pp. 35–87, McGraw-Hill, New York (1961)
25. Kluksdahl, N. C., Kriman, A. M., Ferry, D. K., Ringhofer, C.: Self-consistent study of the resonant tunneling diode. Phys. Rev. B. **39**, 7720–7735 (1989)
26. Lee, H.: Theory and application of the quantum phase-space distribution functions. Phys. Rep. **259**, 147–211 (1995)

27. Lindblad, G.: On the generators of quantum dynamical semigroups. Commun. Math. Phys. **48**, 119–130 (1976)
28. López, J. L.: Nonlinear Ginzburg-Landau-type approach to quantum dissipation. Phys. Rev. E **69**, 026110–026125 (2004)
29. Markowich, P. A.: On the equivalence of the Schrödinger and the quantum Liouville equations. Math. Meth. in the Appl. Sci. **11**, 459–469 (1989)
30. Markowich, P. A., Ringhofer, C. A., Schmeiser, C.: Semiconductor Equations. Springer, Wien (1990)
31. Manfredi, G., Haas, F.: Self-consistent fluid model for a quantum electron gas. Phys. Rev. B **64**, 075316–075323 (2001)
32. Mehta, C. L.: Phase-space formulation of the dynamics of canonical variables. J. Math. Phys. **5**, 677–686 (1964)
33. Santer, M., Mehlig, B., Moseler, M.: Optical response of two-dimensional electron fluids beyond the Kohn regime: strong nonparabolic confinement and intense laser light. Phys. Rev. Lett. **89**, 266801–266804 (2002)
34. Sudarshan, E. C. G.: Equivalence of semiclassical and quantum mechanical descriptions of statistical light beams. Phys. Rev. Lett. **10**, 277–279 (1963)
35. Suh, N., Feix, M. R. and Bertrand, P.: Numerical simulation of the quantum Liouville-Poisson system. J. Comput. Phys. **94**, 403–418 (1991)
36. Tatarskii, V. I.: The Wigner representation of quantum mechanics. Sov. Phys. Usp. **26**, 311–327 (1983)
37. Weyl, H.: Quantenmechanik und gruppentheorie. Z. Phys. **46**, 1–46 (1927)
38. Wigner, E.: On the quantum correction for thermodynamic equilibrium. Phys. Rev. **40**, 749–759 (1932)
39. Yvon, J.: La Théorie Statistique des Fluides. Hermann, Paris (1935)

Chapter 3
The Quantum Two-Stream Instability

Abstract The quantum equivalent of the Dawson multistream model is constructed in terms of the fluid variables representation of the Schrödinger–Poisson system. This Madelung-type hydrodynamic formulation is a first step toward a quantum hydrodynamic model for plasmas. The linear dispersion relation as well as the nonlinear stationary states are discussed, in the one- and two-stream cases. The quantum two-stream instability is analyzed in terms of the coupling of approximate fast and slow waves carrying positive and negative energies.

3.1 Streaming Instabilities in Quantum Plasmas

Since it has been discussed for the first time in the framework of a quantum hydrodynamical model [12], the quantum two-stream instability has attracted considerable attention in the literature. The reason for this is that it is a benchmark displaying many of the particularities of quantum plasmas, including a new unstable branch of the dispersion relation for large wavenumber and almost stationary, quasineutral, nonlinear oscillations [12] without analog in classical plasmas. Furthermore, it has been found [3] that temperature effects can suppress the purely quantum instabilities, as described by a kinetic (Wigner–Poisson) treatment. Similarly, the thermal spread is responsible for a smaller classical two-stream instability treated by the Vlasov–Poisson system. In addition, the quantum fluid equations have been used for several quantum streaming instability problems, like in quantum dusty plasmas [2], in three-stream quantum plasmas [13] or in electron–positron–ion quantum plasmas [16]. The hydrodynamic formalism has also been applied to the quantum filamentation instability, with or without magnetization [8, 9].

F. Haas, *Quantum Plasmas: An Hydrodynamic Approach*, Springer Series on Atomic, Optical, and Plasma Physics 65, DOI 10.1007/978-1-4419-8201-8_3, © Springer Science+Business Media, LLC 2011

3.2   Quantum Dawson Model

In the present chapter, we consider a one-dimensional quantum plasma, where the electrons are described by a statistical mixture of M pure states, each with wavefunction $\psi_\alpha, \alpha = 1, ..., M$ obeying the Schrödinger–Poisson system

$$i\hbar \frac{\partial \psi_\alpha}{\partial t} = -\frac{\hbar^2}{2m} \frac{\partial^2 \psi_\alpha}{\partial x^2} - e\phi \psi_\alpha , \quad \alpha = 1, ..., M, \tag{3.1}$$

$$\frac{\partial^2 \phi}{\partial x^2} = \frac{e}{\varepsilon_0} \left(\sum_{\alpha=1}^{M} |\psi_\alpha|^2 - n_0 \right), \tag{3.2}$$

where $\phi(x,t)$ is the electrostatic potential. Electrons have mass m and charge $-e$, and are globally neutralized by a fixed ion background with density n_0. We assume periodic boundary conditions, with spatial period L. Finally, in the context of this chapter, it is convenient to adopt the normalization

$$\int dx |\psi_\alpha|^2 = N/M, \tag{3.3}$$

where N is the number of particles in a length L so that $n_0 = N/L$ and global charge neutrality is assured.

The system of (3.1) and (3.2) takes into account diffraction, which is the most evident quantum effect, but neglects dissipation, spin and relativistic corrections. Nevertheless, it is useful to consider simplified models that capture the main features of quantum plasmas. Indeed, (3.1) and (3.2) are sufficiently rich to display a wide variety of behaviors, as will be seen soon. At the same time, the model is still amenable to analytic and numerical treatment.

As discussed in Chap. 2, a physically equivalent approach would consist in considering a Wigner function describing the same mixture. However, for analytic and numeric purposes, the Schrödinger–Poisson model reveal to be more convenient. In particular, for the numerical simulations the Wigner formalism is cast into a two-dimensional phase space, whilst the Schrödinger–Poisson model only requires the discretization of an one-dimensional configuration space. Of course, if the number M of streams is large, the numerical cost for the description of the system of (3.1) and (3.2) is also considerable. Nevertheless, interesting physical phenomena (such as instabilities, for $M = 2$) can take place even with a few streams.

For the analytical study, the hydrodynamic formulation of the Schrödinger–Poisson system is particularly convenient, since it makes direct use of macroscopic plasma quantities, such as density and average velocity. Moreover, it enables one to perform straightforward perturbation calculations in the same fashion as in the classical case. Hence, let us introduce the amplitude $A_\alpha = A_\alpha(x,t)$ and the phase $S_\alpha = S_\alpha(x,t)$ associated with the pure state $\psi_\alpha = \psi_\alpha(x,t)$ according to

$$\psi_\alpha = A_\alpha \exp(i S_\alpha/\hbar). \tag{3.4}$$

Both A_α and S_α are defined as real quantities. The density n_α and the velocity u_α of the αth stream of the plasma are given by

$$n_\alpha = A_\alpha^2, \qquad u_\alpha = \frac{1}{m}\frac{\partial S_\alpha}{\partial x}. \tag{3.5}$$

Introducing (3.4) and (3.5) into (3.1) and (3.2) and separating the real and imaginary parts of the equations, we find

$$\frac{\partial n_\alpha}{\partial t} + \frac{\partial}{\partial x}(n_\alpha u_\alpha) = 0, \tag{3.6}$$

$$\frac{\partial u_\alpha}{\partial t} + u_\alpha\frac{\partial u_\alpha}{\partial x} = \frac{e}{m}\frac{\partial \phi}{\partial x} + \frac{\hbar^2}{2m^2}\frac{\partial}{\partial x}\left(\frac{\partial^2(\sqrt{n_\alpha})/\partial x^2}{\sqrt{n_\alpha}}\right), \tag{3.7}$$

$$\frac{\partial^2 \phi}{\partial x^2} = \frac{e}{\varepsilon_0}\left(\sum_{\alpha=1}^{M} n_\alpha - n_0\right). \tag{3.8}$$

The continuity equation (3.6) and the quantum Euler equation (3.7) are the fluid dynamics representation of the Schrödinger equation, as introduced by Madelung [15]. In this context, (3.4) can be termed the Madelung decomposition of the wavefunction.

In the resulting set of equations, quantum effects are contained in the pressure-like, $\hbar$-dependent term in (3.7). If we set $\hbar = 0$, we simply obtain the classical multistream model introduced by Dawson [10]. Therefore, we shall refer to (3.6)–(3.8) as the quantum multistream, or quantum Dawson model [12]. Let us examine the consequences of the quantum Dawson model.

3.3 One-Stream Plasma

To introduce the basic ideas, we first consider the one-stream case and take $M = 1$, that is, a single pure quantum state. For brevity, we write $n_1 \equiv n$, $u_1 \equiv u$. We obtain

$$\frac{\partial n}{\partial t} + \frac{\partial}{\partial x}(nu) = 0, \tag{3.9}$$

$$\frac{\partial u}{\partial t} + u\frac{\partial u}{\partial x} = \frac{e}{m}\frac{\partial \phi}{\partial x} + \frac{\hbar^2}{2m^2}\frac{\partial}{\partial x}\left(\frac{\partial^2(\sqrt{n})/\partial x^2}{\sqrt{n}}\right), \tag{3.10}$$

$$\frac{\partial^2 \phi}{\partial x^2} = \frac{e}{\varepsilon_0}(n - n_0). \tag{3.11}$$

The homogeneous solution for (3.9)–(3.11) is given by

$$n = n_0, \qquad u = u_0, \qquad \phi = 0, \tag{3.12}$$

where u_0 is a constant representing the equilibrium velocity of the stream. The linear stability of this solution is obtained by Fourier analyzing (3.9)–(3.11),

$$n = n_0 + n' \exp(\mathrm{i}(kx - \omega t)), \tag{3.13}$$

$$u = u_0 + u' \exp(\mathrm{i}(kx - \omega t)), \tag{3.14}$$

$$\phi = \phi' \exp(\mathrm{i}(kx - \omega t)) \tag{3.15}$$

Retaining only terms up to first-order in n', u', and ϕ' a linear homogeneous system for the perturbations is found. Nontrivial solutions exist provided the dispersion relation

$$(\omega - ku_0)^2 = \omega_{\mathrm{p}}^2 + \frac{\hbar^2 k^4}{4m^2}, \tag{3.16}$$

holds, where $\omega_{\mathrm{p}} = (n_0 e^2/m\varepsilon_0)^{1/2}$ is the plasma frequency. The zero-temperature Bohm–Pines dispersion relation [7] is recovered, the term ku_0 just representing a Doppler shift. Since the frequency ω is always real, we have undamped stable oscillations of the plasma.

The classical analog of this system is the "cold plasma" model because there's no pressure term in the momentum equation. However, in the quantum realm the Bohm potential term plays a rôle similar to a pressure, even if in mathematical terms it does not correspond to the gradient of a function of the density only. The Bohm term arises directly from the Schrödinger equation, and is responsible for typical quantum-like behavior involving tunneling and wave packet spreading. Formally, it contribute to extra dispersion of the small wavelengths, as apparent from (3.16). This is relevant when we compare the propagation of nonlinear waves in the classical and quantum cases. The classical cold plasma model is known to sustain nonlinear oscillations when the amplitude of the initial perturbation is smaller than a certain value. Beyond this value, the solution becomes singular in a finite time, which is a sign that the model is no longer valid. This phenomenon corresponds to the breaking of the plasma wave, due to particle overtaking in the phase space. On the other hand, due the Bohm pressure-like term in (3.10), the quantum solution never becomes singular, as confirmed by computer simulations [6].

Unlike in the kinetic formalism, the stationary solutions of the fluid equations are fairly amenable to analyze. Defining all quantities to depend only on position, (3.9) and (3.10) reduces to

$$\frac{\mathrm{d}}{\mathrm{d}x}(nu) = 0, \tag{3.17}$$

$$u\frac{\mathrm{d}u}{\mathrm{d}x} = \frac{e}{m}\frac{\mathrm{d}\phi}{\mathrm{d}x} + \frac{\hbar^2}{2m^2}\frac{\mathrm{d}}{\mathrm{d}x}\left(\frac{\mathrm{d}^2(\sqrt{n})/\mathrm{d}x^2}{\sqrt{n}}\right). \tag{3.18}$$

Equations (3.17) and (3.18) have the first integrals

$$J = nu, \tag{3.19}$$

$$\Xi = \frac{mu^2}{2} - e\phi - \frac{\hbar^2}{2m}\left(\frac{\mathrm{d}^2(\sqrt{n})/\mathrm{d}x^2}{\sqrt{n}}\right), \tag{3.20}$$

or

$$\frac{dJ}{dx} = 0, \quad \frac{d\Xi}{dx} = 0, \tag{3.21}$$

corresponding to charge and energy conservation. The constant $\mathscr{H}$ can be eliminated by the global shift $\phi \to \phi + \Xi/e$, and therefore we assume $\Xi = 0$. Then, eliminating u, introducing $A = \sqrt{n}$ and using Poisson's equation, we obtain

$$\hbar^2 \frac{d^2 A}{dx^2} = m\left(\frac{mJ^2}{A^3} - 2eA\phi\right), \tag{3.22}$$

$$\frac{d^2 \phi}{dx^2} = \frac{e}{\varepsilon_0}(A^2 - n_0). \tag{3.23}$$

It can be verified that the $J = 0$ case cannot sustain small-amplitude, periodic solutions. Hence, we assume $J = n_0 u_0$ with $u_0 \neq 0$ and introduce the following rescaling

$$x^* = \frac{\omega_p x}{u_0}, \quad A^* = \frac{A}{\sqrt{n_0}},$$

$$\phi^* = \frac{e\phi}{mu_0^2}, \quad H = \frac{\hbar\omega_p}{mu_0^2}. \tag{3.24}$$

We obtain, in the transformed variables (omitting the stars for simplicity of notation),

$$H^2 \frac{d^2 A}{dx^2} = -2\phi A + \frac{1}{A^3}, \tag{3.25}$$

$$\frac{d^2 \phi}{dx^2} = A^2 - 1, \tag{3.26}$$

a system depending only on the rescaled parameter H, which is the proper measure of the importance of quantum effects. Physically, H is the ratio between the plasmon energy $\hbar\omega_p$ and the kinetic energy mu_0^2 of a particle in the beam.

Let us consider the classical limit of (3.25) and (3.26). This classical limit is singular, because when $H = 0$, (3.25) degenerates into an algebraic equation, yielding $A^2 = \pm 1/\sqrt{2\phi}$. Since the amplitude A is supposed real, we need to choose the positive solution so that the equation for the electrostatic potential becomes

$$\frac{d^2 \phi}{dx^2} = \frac{1}{\sqrt{2\phi}} - 1. \tag{3.27}$$

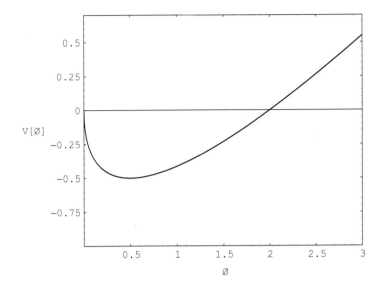

Fig. 3.1 Potential function $V(\phi)$ at (3.28)

Equation (3.27) corresponds to an autonomous, one degree of freedom Hamiltonian problem, for a "particle" moving in a potential $V(\phi) = \phi - \sqrt{2\phi}$:

$$\frac{d^2\phi}{dx^2} = -\frac{dV(\phi)}{d\phi} = -\frac{d}{d\phi}\left[\phi - \sqrt{2\phi}\right]. \tag{3.28}$$

Using this analogy, one can see that (3.27) has periodic solutions around the equilibrium $\phi = 1/2$, with the necessary condition that the initial condition satisfies $0 < \phi(x = 0) < 2$. This can be seen in Fig. 3.1 showing the potential $V(\phi)$. The fact that linear oscillations around $\phi = 1/2$ are sustained is apparent. For nonlinear oscillations with an energy bigger than zero one sees that eventually the oscillation will come back toward $\phi = 0$, implying a complex $V(\phi)$ in a finite time.

The fact that no solution exists for sufficiently large values of the potential is easily understood. A large potential fluctuation induces a velocity fluctuation, which can drive $u(x)$ far from its nominal value u_0. If the potential is sufficiently strong, $u(x)$ can even vanish, but in that case the relation $nu = J = $ constant implies an infinite density. This is the well-known effect of particle overtaking that occurs in the cold plasma model.

Going back to the quantum mechanical case, we have shown that (3.25) and (3.26) describe a quantum fluid version of the Bernstein–Greene–Kruskal (BGK) inhomogeneous equilibria of the one-component plasma. In classical plasma [5] the BGK solution refers to exact nonlinear stationary solutions for the Vlasov–Poisson system where the distribution function is expressed in terms of the energy of a particle under a time-independent scalar potential.

In a nutshell, the BGK method consider the stationary Vlasov's equation

$$v\frac{\partial f}{\partial x} - \frac{e\phi(x)}{m}\frac{\partial f}{\partial v} = 0 \tag{3.29}$$

in the one degree of freedom case, for the particle distribution function $f(x, v)$. From the method of characteristics, we know the general solution for (3.29) to be in the form $f = F(\mathscr{H})$, where F is an arbitrary function of the energy

$$\mathscr{H} = \frac{mv^2}{2} - e\phi(x). \tag{3.30}$$

For a specific form of $F(\mathscr{H})$ one transform Poisson's equation into an autonomous one-dimensional Hamiltonian system for $\phi(x)$, which is known to be integrable by quadrature. However, it is hard to extend the BGK approach to the Wigner–Poisson realm since a general function of the constants of motion is not a solution for the quantum Vlasov equation. However, in the fluid formulation, some analytical results can be obtained for (3.25) and (3.26), shown below.

Equations (3.25) and (3.26) can be put into Hamiltonian form using the variables

$$\bar{A} = iA, \quad \bar{\phi} = \phi/H. \tag{3.31}$$

Notice that the rescaled amplitude $\bar{A}$ is a purely imaginary quantity. We have

$$\frac{d^2\bar{A}}{dx^2} = -\frac{\partial U}{\partial \bar{A}}, \quad \frac{d^2\bar{\phi}}{dx^2} = -\frac{\partial U}{\partial \bar{\phi}}, \tag{3.32}$$

where $U \equiv U(\bar{A}, \bar{\phi})$ is the pseudo-potential

$$U(\bar{A}, \bar{\phi}) = \frac{1}{H}(1 + \bar{A}^2)\bar{\phi} + \frac{1}{2H^2\bar{A}^2}. \tag{3.33}$$

Since the equations of motion are autonomous with respect to the independent variable x, the Hamiltonian formulation immediately gives the first integral

$$I = \frac{1}{2}\left[\left(\frac{d\bar{A}}{dx}\right)^2 + \left(\frac{d\bar{\phi}}{dx^2}\right)^2\right] + U(\bar{A}, \bar{\phi}), \tag{3.34}$$

or

$$\frac{dI}{dx} = 0. \tag{3.35}$$

which is the Hamiltonian function in transformed coordinates. Transforming back to the original variables, one obtains the first integral for (3.25) and (3.26)

$$I = \frac{1}{2}\left[-\left(\frac{dA}{dx}\right)^2 + \frac{1}{H^2}\left(\frac{d\phi}{dx}\right)^2\right] + \frac{1}{H^2}(1 - A^2)\phi - \frac{1}{2H^2A^2}. \tag{3.36}$$

According to the Liouville–Arnold theorem [4], an autonomous two degrees of freedom Hamiltonian system is completely integrable if it possesses two first integrals in involution (null mutual Poisson bracket) and with compact level surfaces. Even if I has not compact level surfaces, a second constant of motion would be an indicative of integrability of the stationary spatial dynamics, restricting the motion to a lower-dimensional manifold. However, no second constant of motion seems to be available for (3.25) and (3.26). Nevertheless, numerical integration for a wide-range of values of H, and different initial conditions, strongly suggest that bounded solutions are always regular. Therefore, an additional hidden first integral probably exist.

It is interesting to perform a linear stability analysis to see in what conditions the system supports small amplitude spatially periodic solutions. Writing

$$A = 1 + A' \exp(ikx), \quad \phi = 1/2 + \phi' \exp(ikx), \tag{3.37}$$

and retaining in (3.25) and (3.26) only terms up to first-order in the primed variables, we obtain the relation

$$H^2 k^4 - 4k^2 + 4 = 0. \tag{3.38}$$

Again, we point out the singular character of the classical limit: for $H = 0$, (3.38) degenerates into a quadratic equation, with solutions $k = \pm 1$. The wavenumber always being real, this corresponds to spatially periodic solutions. When $H \neq 0$, we obtain

$$k^2 = \frac{2 \pm 2\sqrt{1 - H^2}}{H^2}. \tag{3.39}$$

For $H < 1$ (semiclassical regime), both wavenumbers are real, and therefore the system can sustain spatially periodic oscillations. For $H > 1$ (strong quantum effects), the solutions are spatially unstable, and grow exponentially. For $H = 1$, the spectrum is degenerate, since then the quartic equation (3.38) has only two double solutions, $k = \pm\sqrt{2}/H$. The corresponding secular terms imply spatially unstable perturbations, growing linearly with x. In conclusion, small-amplitude stationary solutions of the one-stream Schrödinger–Poisson system can only exist in the semiclassical regime, $H < 1$.

3.4 Two-Stream Plasma

3.4.1 Two Counter Propagating Beams

Classically, the case of two counter propagating beams can have unstable electrostatic oscillations. For the quantum case [12], we consider (3.6)–(3.8) with $M = 2$. Linearizing around the equilibrium

$$n_1 = n_2 = n_0/2, \quad u_1 = -u_2 = u_0, \quad \phi = 0, \tag{3.40}$$

where $u_0 \neq 0$ is the beam speed, and Fourier transform as in the one-stream case. It is convenient to use the dimensionless variables

$$\Omega = \omega/\omega_p, \quad K = ku_0/\omega_p, \quad H = \hbar\omega_p/mu_0^2, \tag{3.41}$$

so that the dispersion relation becomes

$$\Omega^4 - \left(1 + 2K^2 + \frac{H^2K^4}{2}\right)\Omega^2 - K^2\left(1 - \frac{H^2K^2}{4}\right)\left(1 - K^2 + \frac{H^2K^4}{4}\right) = 0. \tag{3.42}$$

This is just a quadratic equation for Ω^2, which can be readily solved as

$$\Omega^2 = \frac{1}{2} + K^2 + \frac{H^2K^4}{4} \pm \frac{1}{2}(1 + 8K^2 + 4H^2K^6)^{1/2}. \tag{3.43}$$

Choosing the positive sign, one always has a positive branch associated with give stable oscillations. Choosing the negative sign in (3.43), one has $\Omega^2 < 0$ provided

$$(H^2K^2 - 4)([HK^2 - 2]^2 + 4K^2[H - 1]) < 0, \tag{3.44}$$

The unstable waves arises through the marginal mode ($\Omega = 0$), since the frequency is either real or purely imaginary. In the classical case ($H = 0$), (3.44) reproduces $K^2 < 1$, which is the classical instability criterion. In the quantum case, (3.44) bifurcates for $H = 1$. If $H > 1$, the second factor is always positive, and the plasma is unstable if $HK < 2$. However, for the semiclassical situation when $H < 1$, there is instability if either

$$0 < H^2K^2 < 2 - 2\sqrt{1 - H^2}, \tag{3.45}$$

or

$$2 + 2\sqrt{1 - H^2} < H^2K^2 < 4, \tag{3.46}$$

as a simple analysis shows. This yields the stability diagram on Fig. 3.2. The lower instability zone is the semiclassical deformation of the classical instability region. We see an increase of the unstable zone as quantum effects are enhanced. The upper instability zone, on the other hand, has no classical analog. The two zones coincide when $H = 1$.

We define K_A, K_B, and K_C as the wavenumbers for which the growth rate vanishes. From (3.45) and (3.46), these wavenumbers are given by

$$H^2K_{A,B}^2 = 2 \pm 2\sqrt{1 - H^2}, \tag{3.47}$$

$$H^2K_C^2 = 4, \tag{3.48}$$

with $K_{A,B}$ associated with the minus (plus) sign, respectively. The following property holds,

$$K_A^2 + K_B^2 = K_C^2. \tag{3.49}$$

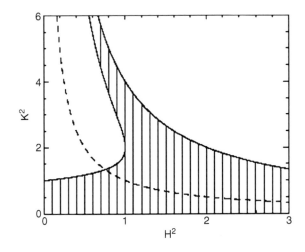

Fig. 3.2 Stability diagram for the two-stream plasma according to [12]. The filled zone is unstable. The *dashed line* corresponds to $HK = 1$, shown for reference. The *lower* and *middle solid curves* correspond to K_A^2 and K_B^2 as defined in (3.47). *Upper solid curve*: K_C^2 as given in (3.48)

These wavelengths are directly related to the stationary solutions of the Schrödinger–Poisson system, as will be verified.

The analysis shows that quantum mechanics has a destabilizing effect in the semiclassical regime, where the unstable zone is bigger than in the classical case. On the other hand, when $H > 2$, fewer modes turn out to be unstable in comparison to the classical plasma case. The results are unexpected, since the quantum mechanics nonlocality can in principle produce stabilization. However, this feature was verified [19] only for large enough H. As will become clear, the stable or unstable character of quantum plasmas depends on subtle properties, with a modified energy transfer between wave modes.

Not only the unstable zone, but also the corresponding growth rate should be determined. In particular, one need to search for the maximum growth rate for a fixed value of H and varying K. Therefore, we define $\Omega = i\gamma$ for the unstable cases with real γ, and plot γ^2 as a function of wavenumber in Fig. 3.3. We see a tendency for the maximum classical growth rate to be larger than the maximum quantum growth rate. By definition, in Fig. 3.3 the intersections with the K axis correspond to wavenumbers K_A, K_B, and K_C as given in (3.47) and (3.48). Notice that the secondary maximum (between wavenumbers K_B and K_C) existent for large wavenumbers is considerably smaller than the first maximum (between $K = 0$ and K_A).

For a truly infinite plasma, the wavenumber (or dimensionless momentum HK) is continuous. However, in a finite system where periodic boundary conditions apply, K is a multiple of the fundamental wavenumber $K_0 = 2\pi/L$, where L correspond to the periodicity length. From an intricate analysis [12], one can show that it is not possible to excite an harmonic in the unstable upper zone of Fig. 3.2, without also exciting the fundamental mode K_0 in the lower unstable region, when $H < 1$.

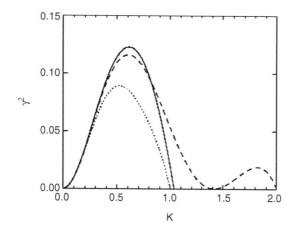

Fig. 3.3 Squared growth rate γ as a function of the wavenumber K, for different values of H, according to [12]. $H = 0.5$, *solid line*; $H = 1$, *dashed line*; $H = 2$, *dotted line*. The intersections of these curves with the K axis correspond to wavenumbers K_A, K_B and K_C as defined in (3.47) and (3.48). For $H = 0.5$, only the intersection at $K_A \simeq 1.035$ is shown; the intersections at $K_B \simeq 3.864$ and $K_C = 4$ are outside the K axis range. For $H = 1$, the intersections are at $K_A = K_B = \sqrt{2}$ and $K_C = 2$. For $H = 2$, there is only one intersection at $K_C = 1$

However, from the same developments, it is found that the fundamental mode is unstable for sufficiently large quantum effects. These results follows from the use of periodic boundary conditions, as for example, in numerical simulations.

3.4.2 Stationary Solutions

After investigating the time-dependent case, it is useful to consider the nonlinear stationary states of (3.6) and (3.7) for $M = 2$. Setting $\partial/\partial t = 0$, one found that the system possess the first integrals

$$J_1 = n_1 u_1, \quad J_2 = n_2 u_2, \tag{3.50}$$

$$E_1 = \frac{m u_1^2}{2} - e\phi - \frac{\hbar^2}{2m} \frac{d^2(\sqrt{n_1})/dx^2}{\sqrt{n_1}}, \tag{3.51}$$

$$E_2 = \frac{m u_2^2}{2} - e\phi - \frac{\hbar^2}{2m} \frac{d^2(\sqrt{n_2})/dx^2}{\sqrt{n_2}}, \tag{3.52}$$

in the sense that $dJ_i/dx = 0, dE_i/dx = 0, i = 1, 2$. Since we are particularly interested in the case of two symmetric streams, we can write

$$J_1 = -J_2 = \frac{n_0 u_0}{2}, \quad E_1 = E_2 = \frac{m u_0^2}{2}, \tag{3.53}$$

where $u_0 \neq 0$ is the beams speed. Defining $n_1 \equiv A_1^2$ and $n_2 \equiv A_2^2$ and the dimensionless variables

$$x^* = \frac{\omega_p x}{u_0}, \quad A_{1,2}^* = \frac{A_{1,2}}{\sqrt{n_0}}, \tag{3.54}$$

$$\phi^* = \frac{(e\phi + E_1)}{mu_0^2}, \quad H = \frac{\hbar \omega_p}{mu_0^2}, \tag{3.55}$$

and eliminating the velocities from (3.50), we found a dynamical system for the densities and the electrostatic potential. Taking into account Poisson's law (3.8) we have (we omit the stars)

$$H^2 \frac{d^2 A_1}{dx^2} = \frac{1}{4A_1^3} - 2\phi A_1, \tag{3.56}$$

$$H^2 \frac{d^2 A_2}{dx^2} = \frac{1}{4A_2^3} - 2\phi A_2, \tag{3.57}$$

$$\frac{d^2 \phi}{dx^2} = A_1^2 + A_2^2 - 1. \tag{3.58}$$

Equations (3.56)–(3.58) constitute a coupled, nonlinear system of three second-order ordinary differential equations, depending on the control parameter H. Notice the singular character of the classical limit $H = 0$, for which (3.56) and (3.57) degenerate to algebraic equations. The nonlinear system for A_1, A_2, and ϕ can be cast into a Hamiltonian form, after a procedure similar to the one employed for the one-stream stationary state equations (3.25) and (3.26). However, the actual expression of the Hamiltonian is rather involved, and not particularly illuminating, so it will be omitted here. Nevertheless, numerical simulations suggest that (3.56)–(3.58) are integrable.

At first, we can expand (3.56)–(3.58) in the vicinity of the spatially homogeneous equilibrium

$$A_1 = A_2 = \frac{1}{\sqrt{2}}, \quad \phi = \frac{1}{2}. \tag{3.59}$$

Supposing perturbations $A_i', \phi' \sim \exp(iKx)$, the following system is obtained,

$$(4 - H^2 K^2)A_i' + \sqrt{2}\phi' = 0, \quad i = 1, 2 \tag{3.60}$$

$$\sqrt{2}(A_1' + A_2') + K^2 \phi' = 0. \tag{3.61}$$

Nontrivial solutions can exist provided

$$(H^2 K^2 - 4)(H^2 K^4 - 4K^2 + 4) = 0, \tag{3.62}$$

This is the same (with an equality sign) as (3.44). Solutions of (3.62) represent wavenumbers for which both the real and the imaginary part of the frequency vanish, and can be considered as the homogeneous limit of generally inhomogeneous stationary states. If $H < 1$ there are three such solutions, which are the wavenumbers K_A, K_B, and K_C defined in (3.47) and (3.48). If $H > 1$, only the solution K_C survives. The other two solutions become complex, so that spatially periodic stationary modes can no longer exist.

Numerical integration of (3.56)–(3.58) confirms the previous results. For instance, it was verified that periodic solutions only exist for $H < 1$. We take $H = 0.7$ and initialize the amplitudes and the potential (at $x = 0$) with their equilibrium value, plus a small perturbation ε, that is, $\phi(0) = 1/2 + \varepsilon_\phi$, $A_i(0) = (1 + \varepsilon_i)/\sqrt{2}$. In agreement with the discussion of the previous paragraph, if we choose $\varepsilon_\phi = 0$ and $\varepsilon_1 = -\varepsilon_2$, the wavenumber $K_C \simeq 2.857$ is linearly excited and thus dominates (Fig. 3.4), while the potential remains very small. On the other hand, if $\varepsilon_1 = \varepsilon_2$ and ε_ϕ is arbitrary, the modes $K_A \simeq 1.08$ and $K_B \simeq 2.645$ are linearly excited (Fig. 3.5). For generic perturbations, all three wavenumbers are excited. Of course these results are strictly valid only for infinitesimally small perturbations. For moderate values, other modes appear (visible on Figs. 3.4 and 3.5, for which $\varepsilon = 0.02$), although the linear wavenumbers are still dominant. For even larger perturbations, bounded solutions no longer exist.

As apparent from the quantum Dawson model, the hydrodynamic formulation of quantum mechanics makes direct use of the physical objects of classical physics (density, velocity, and pressure). Moreover, the stability analysis and perturbation calculations become straightforward in the hydrodynamic formulation. On the other hand, the Schrödinger–Poisson representation is more convenient for the time-dependent simulations, since accurate numerical techniques for the Schrödinger equation are well known from the computational literature. In particular, extensive numerical simulations have shown that quasi-neutral, spatially periodic, stationary states can be created in the two-stream plasma, and can survive over long times [12]. At this point, we remark the easy way to access nonlinear regimes using fluid models, in comparison with kinetic theory, as expressed through the dynamical system (3.56)–(3.58).

The presence of streaming instabilities in quantum plasmas originates from a free energy source and mode coupling, as detailed in the next section.

3.5 Physical Interpretation of the Quantum Two-Stream Instability

We shall provide an intuitive explanation of the quantum two-stream instability, in terms of the coupling of electrostatic modes with distinct energy contents. With this, we offer a physical ("with the hands") understanding of the quantum streaming instabilities. In particular, what is the origin of the quantum unstable modes for large

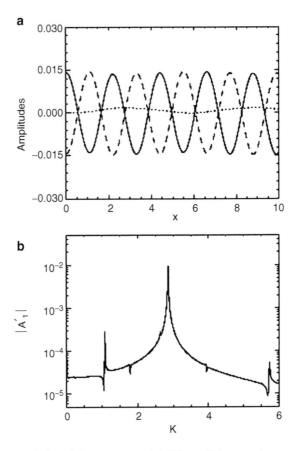

Fig. 3.4 Stationary solution of the two-stream Schrödinger–Poisson system according to [12], with $H = 0.7$, $\varepsilon_\phi = 0$, $\varepsilon_1 = -\varepsilon_2 = 0.02$. (**a**) Spatial variation of the density fluctuations A_1' (*solid line*), A_2' (*dashed line*), and potential fluctuations ϕ' (*dotted line*). Notice that the potential remains small. (**b**) Fourier transform of A_1': the linear wavenumber $K_C \simeq 2.857$ is dominant

wavenumbers, as depicted in Fig. 3.3? In the following, we show how approximate positive and negative energy modes can be identified. The interaction between these waves is the clue for the stability analysis in such systems, as demonstrated in [11].

In classical systems, negative energy modes are a well-known tool for the analysis of streaming instabilities [1, 18]. The heuristic concept of negative energy wave is as follows. Consider, for definiteness, an electromagnetic wave in a dispersive medium into which a beam of particles is also present. When the phase velocity of the wave is slightly smaller than the beam's velocity, on average there can be an energy transfer from the beam to the wave, driving the instability. In this context, the oscillation mode is referred to as a negative energy wave (since it has less energy than the beam). Such simple idea apply to wave propagation in any kind of system, be it classical or quantum.

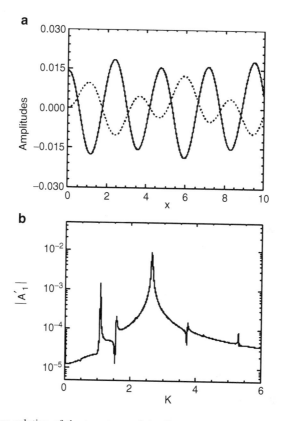

Fig. 3.5 Stationary solution of the two-stream Schrödinger–Poisson system according to [12], with $H = 0.7$, $\varepsilon_\phi = 0$, $\varepsilon_1 = \varepsilon_2 = 0.02$. (**a**) Spatial variation of the density fluctuations A'_1 (*solid line*), A'_2 (*dashed line*), and potential fluctuations ϕ' (*dotted line*). The *solid* and *dashed lines* are superposed, since $A'_1 \simeq A'_2$. (**b**) Fourier transform of A'_1: the linear wavenumbers $K_A \simeq 1.080$ and $K_B \simeq 2.645$ are dominant

3.5.1 Time-Averaged Energy Density of Electrostatic Oscillations

The dispersion relation in (3.43) can be rewritten according to $\Omega^2 = \Omega^2_\pm(K)$, where

$$\Omega_+ = \frac{1}{2}\left[2 + 4K^2 + H^2 K^4 + 2\sqrt{1 + 8K^2 + 4H^2 K^6}\right]^{1/2}, \tag{3.63}$$

$$\Omega_- = \frac{1}{2}\left[2 + 4K^2 + H^2 K^4 - 2\sqrt{1 + 8K^2 + 4H^2 K^6}\right]^{1/2}, \tag{3.64}$$

associated with four possible branches for the eigen-frequency Ω as a function of the wavenumber K. We use parity properties to restrict the analysis to positive K and Ω values. As detailed in Sect. 3.4.1, when $0 < H < 1$, there is instability provided

$K < K_A$ (semiclassical branch) or $K_B < K < K_C$ (quantum branch), where $K_A < K_B < K_C$ are given by (3.47), or

$$
K_A = \frac{\left[2 - 2\sqrt{1 - H^2}\,\right]^{1/2}}{H}, \quad K_B = \frac{\left[2 + 2\sqrt{1 - H^2}\,\right]^{1/2}}{H}, \quad K_C = \frac{2}{H}. \tag{3.65}
$$

On the other hand, when $H \geq 1$ the instability condition is just $K < K_C$ (see Fig. 3.2).

To evaluate the energy content of a wave mode, it is necessary to consider the time-averaged energy density $\langle W_e \rangle$ of electrostatic oscillations [17], which turns out to be

$$
\langle W_e \rangle = \frac{\varepsilon_0}{4} \frac{\partial (\Omega \, \varepsilon_h)}{\partial \Omega} |E_1|^2, \tag{3.66}
$$

where ε_0 is the vacuum permittivity, ε_h is the Hermitian part of the dielectric function and E_1 is the amplitude of the perturbation electric field. The dielectric function ε is defined so that after linearizing the fluid equations one has

$$
\varepsilon E_1 = 0. \tag{3.67}
$$

For the symmetric two-stream case it is

$$
\varepsilon = 1 - \frac{1}{2} \left[\frac{1}{(\Omega + K)^2 - H^2 K^4/4} + \frac{1}{(\Omega - K)^2 - H^2 K^4/4} \right]. \tag{3.68}
$$

Since there is no dissipation mechanism in the present model, one has $\varepsilon = \varepsilon_h$, as is evident since the dielectric function is real.

Also notice that the derivation of (3.66) relies on Maxwell's equations only [17], so that it apply to quantum plasmas too. The difference to classical physics is that quantum effects are present in the modified dielectric function. In addition, (3.66) can be alternatively found from a generalized Poynting theorem, in a similar way as for the classical two-stream instability [14].

Proceeding from (3.66) and (3.68), one get

$$
\partial (\Omega \, \varepsilon_h)/\partial \Omega \sim \psi(\Omega), \tag{3.69}
$$

omitting a complicated positive factor, where

$$
\psi(\Omega) \equiv -6K^4 + H^2 K^6 + \frac{H^2 K^8}{8} + K^2 (4 - H^2 K^2) \Omega^2 + 2\Omega^4. \tag{3.70}
$$

Evaluating for the wave mode $\Omega = \Omega_+(K)$ using (3.63), the result is

$$
\psi(\Omega_+) = 1 + 8K^2 + 4H^2 K^6 + (1 + 4K^2) \sqrt{1 + 8K^2 + 4H^2 K^6} > 0, \tag{3.71}
$$

so that this is always a *positive energy* wave.

On the other hand, evaluating for the wave mode $\Omega = \Omega_-(K)$ using (3.64), the result is

$$\psi(\Omega_-) = 1 + 8K^2 + 4H^2 K^6 - (1 + 4K^2)\sqrt{1 + 8K^2 + 4H^2 K^6}. \qquad (3.72)$$

which has a negative contribution. From (3.65) and (3.72), it can be shown that $\psi(\Omega_-) < 0$ if and only if $K < K_C$. Therefore, if $K > K_C$, the mode $\Omega = \Omega_-(K)$ is a stable *positive energy* mode; if $K < K_C$, the unstable modes described before (3.65) and by $\Omega = \Omega_-(K)$ are *negative energy* waves. Actually, they correspond to absolute instability in the sense that they are of the form $\Omega = i\gamma$, for real $\gamma > 0$. In addition, notice that the stable mode $\Omega = \Omega_-(K)$ for $K_A < K < K_B$, existing only for $H < 1$, carry negative energy. Finally, $\psi(\Omega_-) = 0$ for the marginal stable wavenumber $K = K_C$, so that this wave carry zero energy.

Further insight can be gained analyzing the characteristic function $F(\Omega)$ such that the dispersion relation for (3.6)–(3.8) in the symmetric two-stream case is expressed as

$$\varepsilon = 1 - F(\Omega) = 0. \qquad (3.73)$$

Here, the dielectric function ε is given in terms of

$$F(\Omega) = \frac{1}{2}\left[\frac{1}{(\Omega + K)^2 - H^2 K^4/4} + \frac{1}{(\Omega - K)^2 - H^2 K^4/4}\right]. \qquad (3.74)$$

The characteristic function has vertical asymptotes at $\Omega = \pm\Omega_>$ and $\Omega = \pm\Omega_<$, where

$$\Omega_> = K + \frac{HK^2}{2}, \quad \Omega_< = K - \frac{HK^2}{2}. \qquad (3.75)$$

Since the dispersion relation (3.42) is a quadratic equation with real coefficients for Ω^2, stability is assured when the graph of $F(\Omega)$ intercept four times the horizontal line $F = 1$. Actually, the case $K = K_C$ is special because the quartic equation for Ω degenerate into a quadratic one, which can be shown to correspond always to stable oscillations. Figure 3.6 shows a typical unstable case when $K < K_C$, for $K = 1$, $H = 0.8$.

On the other hand, Fig. 3.7 gives insight into why the wavenumbers satisfying $K > K_C$ are stable, since the graph of the characteristic function always intercept the horizontal line $F = 1$ four times.

3.5.2 Fast and Slow Approximate Modes in Electrostatic Two-Stream Quantum Plasmas

Figure 3.8 shows the dispersion curves for $H < 1$. We take only real wavenumbers, so that the amplification problem is not considered. In other words, the alternative possibility of spatially growing solutions for imaginary K is not analyzed.

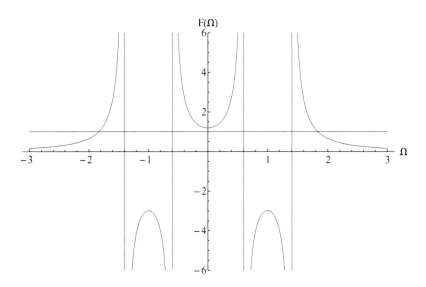

Fig. 3.6 Characteristic function for $K < K_C$ according to [11]. In the example, $K = 1, H = 0.8$. It corresponds to instability, since $F(0) > 1$

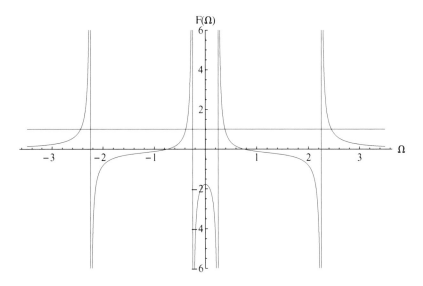

Fig. 3.7 Characteristic function for the stable wavenumbers $K > K_C$ according to [11]. In the example, $K = 1, H = 2.5$

In Fig. 3.8, the curve 1 is a positive energy mode parametrized by $\Omega = \Omega_+(K)$ given by (3.63). Curves 2 and 3 are both described by $\Omega = \Omega_-(K)$ given by (3.64). However, curve 2 carry negative energy while curve 3 is a positive energy mode. The coupling of these waves gives rise to the purely quantum (absolute) instability for large wavenumbers, $K_B < K < K_C$ in Fig. 3.8.

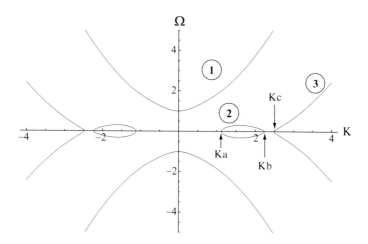

Fig. 3.8 Dispersion curves for $H = 0.8$ according to [11]. Curve 1 is a positive energy mode parametrized by $\Omega = \Omega_+(K)$. Curve 2 is a negative energy mode, while curve 3 is a positive energy mode. Both curves 2 and 3 are described by $\Omega = \Omega_-(K)$, see (3.63) and (3.64)

However, the referred coupling is not exact since curves 2 and 3 did not touch in Fig. 3.8. To give support to the interpretation, we follow the style of [14] and write the dispersion relation in the factorized form

$$
\left[\Omega - K - \frac{1}{\sqrt{2}}\left(1 + \frac{H^2 K^4}{2}\right)^{1/2}\right]\left[\Omega - K + \frac{1}{\sqrt{2}}\left(1 + \frac{H^2 K^4}{2}\right)^{1/2}\right]
$$

$$
\times \left[\Omega + K - \frac{1}{\sqrt{2}}\left(1 + \frac{H^2 K^4}{2}\right)^{1/2}\right]\left[\Omega + K + \frac{1}{\sqrt{2}}\left(1 + \frac{H^2 K^4}{2}\right)^{1/2}\right] = \frac{1}{4}.
$$

$$(3.76)$$

Hence, two fast

$$
\Omega \simeq \Omega_f \equiv \pm K + \frac{1}{\sqrt{2}}\left(1 + \frac{H^2 K^4}{2}\right)^{1/2}
\tag{3.77}
$$

and two slow

$$
\Omega \simeq \Omega_s \equiv \pm K - \frac{1}{\sqrt{2}}\left(1 + \frac{H^2 K^4}{2}\right)^{1/2}
\tag{3.78}
$$

approximate space-charge modes can be identified. The sign of the linear in K term correspond to Doppler-shifted quantum Langmuir waves associated with the positive or negative propagating electron beams. Paying attention to the beam propagating in the positive direction and using (3.63) and (3.64), it can be seen that

$$
\Omega_+ \simeq \Omega_f \simeq K + \frac{H K^2}{2}, \quad \Omega_- \simeq \Omega_s \simeq K - \frac{H K^2}{2}
\tag{3.79}
$$

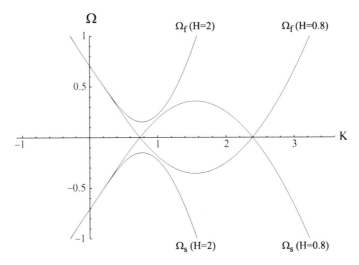

Fig. 3.9 Fast $\Omega_f = -K + (1 + H^2 K^4/2)^{1/2}/\sqrt{2}$ and slow $\Omega_s = K - (1 + H^2 K^4/2)^{1/2}/\sqrt{2}$ asymptotic modes for $H = 0.8$ and $H = 2$ according to [11]

for large K, so that the fast and slow waves in (3.77) and (3.78) are the asymptotic forms of the four exact branches of the dispersion relation.

Calculating the time-averaged energy density $\langle W_e \rangle$ of electrostatic oscillations using (3.66), it can be directly verified that the fast space-charge waves in (3.77) always have positive energy. On the other hand, it is found that the slow modes in (3.78) are negative energy waves provided $K < K_C$, since they correspond to the exact waves $\Omega = \Omega_-(K)$.

Instability is expected when the fast positive energy wave of the beam propagating in the negative direction couples to the slow negative energy wave propagating in the positive direction, or vice versa. For positive (Ω, K), one then need to put $\Omega_f = \Omega_s$, choosing the negative sign in (3.77) and the positive sign in (3.78). Proceeding in a similar way allowing also for negative wavenumbers gives the coupling condition

$$\frac{1}{\sqrt{2}}\left(1 + \frac{H^2 K^4}{2}\right)^{1/2} = \pm K, \tag{3.80}$$

as illustrated in Fig. 3.9 showing the intersection of the fast and slow modes for $H = 0.8$. The coupling occurs for $\Omega_{f,s} = 0$, in accordance with the fact that instability occurs for zero real part of the frequency. Hence, ignoring the factor $1/4$ on the right-hand side of (3.76), we discover an interaction between a fast positive wave and a slow negative wave, as apparent from the crossing of the curves.

In the context of this interpretation, the wavenumbers K_m satisfying the coupling condition (3.80) correspond to maximal instability growth rate. Solving for K^2 for $H \neq 0$, these wavenumbers are found to be given by

$$K_m^2 = \frac{2}{H^2}\left[1 \pm \sqrt{1 - \frac{H^2}{2}}\right].$$

(3.81)

Notice that in the classical case, where $H = 0$, one would have from (3.80) only the solution $K_m^2 = 1/2$.

Taking the plus sign in (3.81), one has

$$K_m \equiv K_{m,q} = \frac{\sqrt{2}}{H}\left[1 + \sqrt{1 - \frac{H^2}{2}}\right]^{1/2}.$$

(3.82)

Assuming $H < 1$, we get, to leading order,

$$K_{m,q} = K_C - \frac{H}{8} + O(H^3) > K_B \simeq K_C - \frac{H}{4} + O(H^3),$$

(3.83)

where K_B and K_C are defined in (3.65). Therefore, $K_B < K_{m,q} < K_C$, which is exactly what should be expected about an instability arising from the coupling of the positive energy mode shown in curve 3 and the negative energy mode shown in curve 2 of Fig. 3.8. This explains the physical origin of the purely quantum instability which can occur for large wavenumbers, when $H < 1$. Comparison between the exact wavenumber for maximal quantum instability and $K_{m,q}$ also shows satisfactory agreement, for a fixed value of $H < 1$. The discrepancy follows since, after all, $K_{m,s}$ are just approximate modes.

On the other hand, taking the minus sign in (3.81), one has

$$K_m \equiv K_{m,c} = \frac{\sqrt{2}}{H}\left[1 - \sqrt{1 - \frac{H^2}{2}}\right]^{1/2}.$$

(3.84)

We get, to leading order,

$$K_{m,c} = \frac{1}{\sqrt{2}} + \frac{H^2}{16\sqrt{2}} + O(H^4) < K_A \simeq 1 + \frac{H^2}{8} + O(H^4),$$

(3.85)

where K_A is defined in (3.65). Properly, the wavenumber $K_{m,c}$ can be referred as the semiclassical branch, since it corresponds to the exact classical wavenumber for maximal instability, $K_c = 1/\sqrt{2}$. Moreover, $K_{m,c} < K_A$ corresponds to the coupling of the positive (curve 1) and negative energy (curve 2) branches in Fig. 3.8. Finally, we found a satisfactory agreement between the exact and approximate values of the wavenumber for maximal growth rate, at a fixed H.

When $H \geq 1$, the elliptic-like branch of Fig. 3.8 disappears and one has the dispersion curves shown in Fig. 3.10, where $H = 1$. As shown from (3.72), both exact branches 1 (described by $\Omega = \Omega_+(K)$) and 2 (described by $\Omega = \Omega_-(K)$) in Fig. 3.10 are positive energy modes. This is a signature of a stabilizing influence of

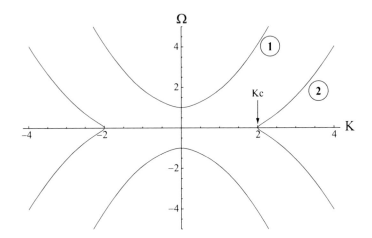

Fig. 3.10 Dispersion curves for $H = 1$ according to [11]. The elliptic-like branch of Fig. 3.8 disappears. Both exact branches 1 (described by $\Omega = \Omega_+(K)$) and 2 (described by $\Omega = \Omega_-(K)$) are positive energy modes

the quantum effects. However, in spite of the existence of only exact positive energy waves, there is an instability for $H \geq 1$, $K < K_C = 2/H$. The situation becomes even more puzzling for large quantum parameters $H > \sqrt{2}$ in (3.82), implying a complex K_m. It happens since the fast Ω_f and slow Ω_s curves have not intersection points for large quantum effects, as seen in Fig. 3.9 for $H = 2$.

To physically understand the origin of the two-stream instability for large quantum parameter, assume a power-law dependence $K = 1/H < K_C$. The linear dispersion relation predicts instability. Calculating the electrostatic energy density W_e from (3.66) for the approximate slow space-charge wave $\Omega_s = K - (1/2)(1 + H^2K^4/2)^{1/2}$ defined in (3.78), one has $W_e \to -\infty$ as $H \to 2$, and $W_e \to +\infty$ as $H \to 3.46$ as shown in Fig. 3.11. The divergences are due to the inexact nature of the slow wave. Nevertheless, the relevant point is that this somewhat hidden approximate mode carry negative energy. Moreover, when $H = 2$, corresponding to the maximal negative electrostatic energy density, one has $1/H = 0.50$. This is near the wavenumber $K = 0.52$ for maximal instability growth rate, for the chosen quantum parameter, as seen in Fig. 3.12. Repeating the procedure considering the fast space-charge modes, one always find a positive energy. Hence, once again there is the coupling between positive and negative waves, giving rise to instability, also for large H.

Assuming a different power-law expression, $K = \alpha/H$, where the parameter α is not necessarily unity, the results are as follows. For $0 < \alpha < 2$, one always has $W_e \to \pm\infty$ for the slow space-charge mode, as $H \to H_\pm > 1$. Hence, there is the possibility of a negative energy wave, for some parameter $H \simeq H_- > 1$. On the other hand, when $\alpha \geq 2$, it can be verified that the slow space-charge mode always carry positive energy. Actually, it diverges to plus infinity, for some value $H \to H_+$. This is not surprising, since the wavenumbers $K > K_C$ are known to be stable.

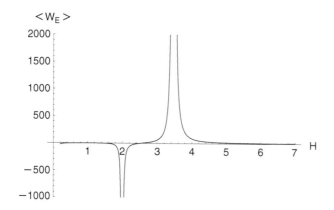

Fig. 3.11 Time-averaged electric energy density $\langle W_e \rangle$ defined in (3.66), evaluated for the slow mode $\Omega = \Omega_s = K - (1/2)(1 + H^2 K^4/2)^{1/2}$ and the unstable wavenumbers $K = 1/H$, apart from the positive factor $\varepsilon_0 |E_1|^2/4$ according to [11]

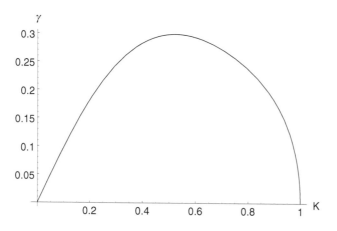

Fig. 3.12 Exact growth rate γ in units of ω_p for $H = 2$, corresponding to the maximal negative energy density waves ($\langle W_E \rangle \rightarrow -\infty$) of Fig. 3.11 for $K = 1/H$ according to [11]. One has instability for $K < K_C = 1$ in this case, and maximal growth rate for $K = 0.52$, close to $K = 1/2$ of Fig. 3.11

The above analysis can in principle be carried on in similar problems, like for the quantum beam-plasma instability or the case of parallel propagating quantum beams.

Problems

3.1. Derive the linear dispersion relation (3.16).

3.2. Perform the numerical simulation of (3.25) and (3.26), analyzing the integrable or nonintegrable features of the system.

3.3. Prove that the current and energy-like quantities in (3.19) and (3.20) are constants of motion along the trajectories of (3.17) and (3.18), or $dJ/dx = 0$, $dE/dx = 0$.

3.4. Verify that (3.17) and (3.18) can not sustain small-amplitude periodic solution in the zero-current $J = 0$ case.

3.5. Directly prove that I in (3.36) is a first integral for the one-stream dynamical system shown in (3.25) and (3.26).

3.6. Derive the dispersion relation (3.39).

3.7. Prove the instability conditions (3.45) and (3.46) for the quantum two-stream instability.

3.8. Find a Hamiltonian formulation (and hence an energy first integral) for (3.56)–(3.58).

3.9. Find the missing positive factor in (3.70) for the function $\psi(\Omega)$ deciding on the energy content of electrostatic waves in the quantum two-stream instability.

3.10. Check (3.71) and (3.72).

3.11. Prove the factorization in (3.76), allowing for the identification of fast and slow approximate space-charge waves in the quantum two-stream problem.

References

1. Akhiezer, A. I., Akhiezer, I. A., Polovin, R. V., Sitenko, A. G. and Stepanov, K. N.: Plasma Electrodynamics: Linear Theory. Pergamon Press, Oxford (1975)
2. Ali, S., Shukla, P. K.: Streaming instability in quantum dusty plasmas. Eur. Phys. J. D **41**, 319–324 (2007)
3. Anderson, D., Hall, B., Lisak, M., Marklund, M.: Statistical effects in the multistream model for quantum plasmas. Phys. Rev. E **65**, 046417–046421 (2002)
4. Arnold, V. I.: Mathematical Methods of Classical Mechanics. Springer, New York (1978)
5. Bernstein, I. B., Greene, J. M., Kruskal, M. D.: Exact nonlinear plasma oscillations. Phys. Rev. **108**, 546–550 (1957)
6. Bertrand, P., Nguyen, V. T., Gros, M., Izrar, B., Feix, M. R., Gutierrez, J.: Classical Vlasov plasma description through quantum numerical methods. J. Plasma Phys. **23**, 401–422 (1980)
7. Bohm, D., Pines, D.: A collective description of electron interactions: III. Coulomb interactions in a degenerate electron gas. Phys. Rev. **92**, 609–625 (1953)
8. Bret, A.: Filamentation instability in a quantum plasma. Phys. Plasmas **14**, 084503–084506 (2007)
9. Bret, A.: Filamentation instability in a quantum magnetized plasma. Phys. Plasmas **15**, 022109–022114 (2008)
10. Dawson, J.: On Landau damping. Phys. Fluids **4**, 869–874 (1961)
11. Haas, F., Bret, A. and Shukla, P. K.: Physical interpretation of the quantum two-stream instability. Phys. Rev. E **80**, 066407–066412 (2009)
12. Haas, F., Manfredi, G., Feix, M.: Multistream model for quantum plasmas. Phys. Rev. E **62**, 2763–2772 (2000)

13. Haas, F., Manfredi, G., Goedert, J.: Stability analysis of a three-stream quantum-plasma equilibrium. Braz. J. Phys. **33**, 128–132 (2003)
14. Lashmore-Davies, C. N.: Two-stream instability, wave energy, and the energy principle. Phys. of Plasmas **14**, 092101–092105 (2007)
15. Madelung, E.: Quantum theory in hydrodynamical form. Z. Phys. **40**, 332–336 (1926)
16. Mushtaq, A., Khan, R.: Linear and nonlinear studies of two-stream instabilities in electron-positron-ion plasmas with quantum corrections. Phys. Scr. **78**, 015501–015505 (2008)
17. Stix, T. H.: Waves in Plasmas. Springer, New York (1992)
18. Sturrock, P. A.: Excitation of plasma oscillations. Phys. Rev. **117**, 1426–1429 (1960)
19. Suh, N. D., Feix, M. R., Bertrand, P.: Numerical simulation of the quantum Liouville-Poisson system. J. Comput Phys. **94**, 403–418 (1991)

Chapter 4
A Fluid Model for Quantum Plasmas

Abstract A quantum fluid model is derived from the Wigner–Poisson system. Quantum statistical effects can be incorporated using a convenient equation of state. Quantum diffraction effects manifest through a Bohm potential term. The derivation is based on the Madelung representation of the ensemble wavefunctions, so that the second-order moment of the Wigner function appear as the sum of kinetic and osmotic pressures and the Bohm potential. The case of an one-dimensional zero-temperature Fermi gas is treated, for both one and two-stream plasmas. The validity conditions for the quantum hydrodynamic model for plasmas are discussed. The derivation of the equation of state for a zero-temperature Fermi gas is detailed for one, two, and three spatial dimensions. The long wavelength condition to avoid kinetic effects is treated in the case of a degenerate plasma. The question of the representation of a given Wigner function in terms of a set of ensemble wavefunctions is worked out.

4.1 The Convenience of Macroscopic Models for Quantum Plasmas

Understanding the dynamics of a quantum electron gas is an important issue for a variety of physical systems, such as ordinary metals, semiconductors, and even astrophysical systems under extreme conditions (e.g., white dwarfs). Although some level of understanding can be achieved by considering independent electrons, a more accurate description requires the use of self-consistent models, where electron–electron interactions are taken into account. As the treatment of the full N-body problem is clearly out of reach, mean field models are usually adopted, of which the Wigner–Poisson and Schrödinger–Poisson systems are examples.

Despite its considerable interest, the Wigner–Poisson formulation presents some intrinsic drawbacks: (a) it is a nonlocal, integro-differential system; and (b) its numerical treatment requires the discretization of the whole phase space. Moreover,

F. Haas, *Quantum Plasmas: An Hydrodynamic Approach*, Springer Series on Atomic, Optical, and Plasma Physics 65, DOI 10.1007/978-1-4419-8201-8_4,

as is often the case with kinetic models, the Wigner–Poisson system gives more information than one is really interested in.

In a similar way, the use of the Schrödinger–Poisson or multistream model is not free of ambiguities. Indeed, the physical interpretation of each wavefunction in the quantum statistical ensemble is not always clear, except in the case of a collection of beams streaming along the plasma. In addition, how to efficiently decompose more general equilibrium distribution functions in terms of a given set of wavefunctions can be a delicate question.

For these reasons, it would be useful to obtain an accurate reduced model which, though not providing the same detailed information as the kinetic Wigner–Poisson system or the equivalent Schrödinger–Poisson system, is still able to reproduce the salient features of quantum plasma systems.

To obtain a set of macroscopic equations for quantum plasmas, we will first derive a system of reduced "fluid" equations by taking moments of the Wigner–Poisson system. Using a Madelung (or eikonal) decomposition, it will be shown that the pressure term appearing in the fluid equations can be separated into a classical and a quantum part. With a working hypothesis about the pressure term, the fluid system can be closed. Moreover, in the quasi-neutral limit, we will derive an effective Schrödinger–Poisson system, which in an appropriate limit, reproduces the results of the kinetic Wigner–Poisson formulation. In this effective Schrödinger–Poisson model, the Schrödinger equation is nonlinear, as it includes an effective potential depending on the modulus of the wavefunction. The exact form of this effective potential depends on the specific physical system being studied. The theory will be applied to a degenerate Fermi gas, including linear wave propagation, nonlinear stationary solutions, and the two-stream instability.

The quantum hydrodynamic model for plasmas was introduced in [20]. Later, the same methodology has been applied to a multitude of problems involving charged particle systems, for instance, the excitation of electrostatic wake fields in nanowires [1], the nonlinear electron dynamics in thin metal films [8], parametric amplification characteristics in piezoelectric semiconductors [12], breather waves in semiconductor quantum wells [15], multidimensional dissipation-based Schrodinger models from quantum Fokker–Planck dynamics [17], the description of quantum diodes in degenerate plasmas [22] and quantum ion-acoustic waves in single-walled carbon nanotubes [24].

4.2 Quantum Fluid Model

For convenience, we rewrite the Wigner–Poisson system given by (2.49) and (2.51),

$$\frac{\partial f}{\partial t} + v\frac{\partial f}{\partial x} - \frac{iem}{2\pi\hbar^2}\int ds\, dv' e^{im(v'-v)s/\hbar}\left[\phi\left(x+\frac{s}{2}\right) - \phi\left(x-\frac{s}{2}\right)\right]f(x,v',t) = 0,$$

$$(4.1)$$

$$\frac{\partial^2 \phi}{\partial x^2} = \frac{e}{\varepsilon_0} \left(\int f \, dv - n_0 \right),$$ (4.2)

where $f(x,v,t)$ is the Wigner distribution function, $\phi(x,t)$ the electrostatic potential, $-e$ and m the electron charge and mass, ε_0 the vacuum dielectric constant and n_0 a background ionic charge. For simplicity of notation, at first only one-dimensional problems will be treated, but the results can be readily extended to higher dimensions.

To derive a fluid model [20], we take moments of (4.1) by integrating over velocity space. Introducing the standard definitions of density, mean velocity and pressure

$$n(x,t) = \int f \, dv, \quad u(x,t) = \frac{1}{n} \int f v \, dv, \quad P(x,t) = m \left(\int f v^2 \, dv - n u^2 \right),$$ (4.3)

it is obtained

$$\frac{\partial n}{\partial t} + \frac{\partial (nu)}{\partial x} = 0,$$ (4.4)

$$\frac{\partial u}{\partial t} + u \frac{\partial u}{\partial x} = \frac{e}{m} \frac{\partial \phi}{\partial x} - \frac{1}{mn} \frac{\partial P}{\partial x}.$$ (4.5)

The continuity equation follows immediately because integrating (4.1) on v eliminate the nonlocal term. For the derivation of (4.5), we multiply (4.1) by v and integrate over velocities using the identity

$$v \exp\left(-\frac{imvs}{\hbar} \right) = \frac{i\hbar}{m} \frac{\partial}{\partial s} \exp\left(-\frac{imvs}{\hbar} \right),$$ (4.6)

taking into account the continuity equation to eliminate $\partial n / \partial t$. A more detailed theory would include the energy transport equation obtained after taking the second-order moment of the Wigner function and the associated time-derivative. However, intriguing new results can be discovered already if we content ourselves with the continuity and force equations.

We immediately notice that (4.4) and (4.5) do not differ from the ordinary evolution equations for a classical fluid. This may seem surprising, but in the following it will appear that the quantum nature of this system is in fact hidden in the pressure term. Contributions where $\hbar$ explicitly appear can be found only in the higher-order moments.

The pressure term may be decomposed into a classical and a quantum part, as follows. The Wigner distribution for a quantum mixture of states $\psi_\alpha(x,t)$, each characterized by an occupation probability $p_\alpha, \alpha = 1,...,M$, is written as

$$f(x,v,t) = \frac{Nm}{2\pi\hbar} \sum_{\alpha=1}^{M} p_\alpha \int ds \, e^{\frac{imvs}{\hbar}} \psi_\alpha^* \left(x + \frac{s}{2} \right) \psi_\alpha \left(x - \frac{s}{2}, t \right),$$ (4.7)

where the sum extends over all possible states. The numbers p_α, representing probabilities, satisfy the relations $p_\alpha \geq 0$, $\sum_{\alpha=1}^{M} p_\alpha = 1$. In terms of the ensemble wavefunctions, from (4.3) one obtains

$$n = N \sum_{\alpha=1}^{M} p_\alpha |\psi_\alpha|^2, \tag{4.8}$$

$$nu = \frac{i\hbar N}{2m} \sum_{\alpha=1}^{M} p_\alpha \left(\psi_\alpha \frac{\partial \psi_\alpha^*}{\partial x} - \psi_\alpha^* \frac{\partial \psi_\alpha}{\partial x} \right), \tag{4.9}$$

and, after some algebra,

$$P = \frac{N\hbar^2}{4m} \sum_{\alpha=1}^{M} p_\alpha \left(2 \left| \frac{\partial \psi_\alpha}{\partial x} \right|^2 - \psi_\alpha^* \frac{\partial^2 \psi_\alpha}{\partial x^2} - \psi_\alpha \frac{\partial^2 \psi_\alpha^*}{\partial x^2} \right)$$

$$+ \frac{N^2\hbar^2}{4mn} \left[\sum_{\alpha=1}^{M} p_\alpha \left(\psi_\alpha^* \frac{\partial \psi_\alpha}{\partial x} - \psi_\alpha \frac{\partial \psi_\alpha^*}{\partial x} \right) \right]^2. \tag{4.10}$$

If we represent each state according to the Madelung [18] decomposition

$$\psi_\alpha(x,t) = A_\alpha(x,t) \exp\left(iS_\alpha(x,t)/\hbar\right), \tag{4.11}$$

where A_α (amplitude) and S_α (phase) are real functions, we get

$$n = N \sum_{\alpha=1}^{M} p_\alpha A_\alpha^2, \tag{4.12}$$

$$nu = \frac{N}{m} \sum_{\alpha=1}^{M} p_\alpha A_\alpha^2 \frac{\partial S_\alpha}{\partial x} \tag{4.13}$$

and also

$$P = \frac{N^2}{2mn} \sum_{\alpha,\beta=1}^{M} p_\alpha p_\beta A_\alpha^2 A_\beta^2 \left(\frac{\partial S_\alpha}{\partial x} - \frac{\partial S_\beta}{\partial x} \right)^2$$

$$+ \frac{N\hbar^2}{2m} \sum_{\alpha=1}^{M} p_\alpha \left[\left(\frac{\partial A_\alpha}{\partial x} \right)^2 - A_\alpha \frac{\partial^2 A_\alpha}{\partial x^2} \right]. \tag{4.14}$$

In the pressure, there is now the explicit presence of $\hbar$. However, since the ensemble wavefunctions satisfy the one-body Schrödinger equation (2.61) both the amplitude and phase implicitly depend on Planck's constant.

At this point, it is useful to define the kinetic u_α and osmotic u_α^0 velocities associated with the wavefunction ψ_α,

$$u_\alpha = \frac{1}{m} \frac{\partial S_\alpha}{\partial x}, \quad u_\alpha^0 = \frac{\hbar}{m} \frac{\partial A_\alpha/\partial x}{A_\alpha}. \tag{4.15}$$

In this way, it can be directly verified that the pressure in (4.14) can be written as

$$P = P^k + P^o + P^Q, \tag{4.16}$$

where the kinetic pressure is

$$P^k = \frac{mn}{2} \sum_{\alpha,\beta=1}^{M} \tilde{p}_\alpha \tilde{p}_\beta (u_\alpha - u_\beta)^2, \tag{4.17}$$

the osmotic pressure is

$$P^o = \frac{mn}{2} \sum_{\alpha,\beta=1}^{M} \tilde{p}_\alpha \tilde{p}_\beta (u_\alpha^o - u_\beta^o)^2, \tag{4.18}$$

and the quantum pressure is

$$P^Q = -\frac{\hbar^2 n}{4m} \frac{\partial^2}{\partial x^2} \ln n. \tag{4.19}$$

In (4.17) and (4.18), a modified set of ensemble probabilities $\tilde{p}_\alpha = \tilde{p}_\alpha(x,t)$ was employed,

$$\tilde{p}_\alpha = \frac{N p_\alpha A_\alpha^2}{n}. \tag{4.20}$$

The new statistical weights satisfy $\tilde{p}_\alpha \geq 0, \sum_{\alpha=1}^{M} \tilde{p}_\alpha = 1$ as they should.

The term "osmotic" is used because the u_α^o are related to density gradients [5,11]. Moreover, for a particular ψ_α, the osmotic velocity points to the regions of higher density, as becomes more evident in the three-dimensional version,

$$\mathbf{u}_\alpha^o = (\hbar/m)\nabla \ln A_\alpha. \tag{4.21}$$

Both pressures P^k and P^o can be viewed as a measure of velocity dispersion. Indeed, consider the following average $\langle f_\alpha \rangle$ of an ensemble function $f_\alpha = f_\alpha(x,t)$:

$$\langle f_\alpha \rangle = \sum_{\alpha=1}^{M} \tilde{p}_\alpha f_\alpha. \tag{4.22}$$

From this definition, it can be easily shown that (4.17) and (4.18) are equivalent to

$$P^k = mn(\langle u_\alpha^2 \rangle - \langle u_\alpha \rangle^2), \tag{4.23}$$

$$P^o = mn(\langle [u_\alpha^o]^2 \rangle - \langle u_\alpha^o \rangle^2). \tag{4.24}$$

For a pure state (so that $\tilde{p}_\alpha = \delta_{\alpha\beta}$ for some β), both P^k and P^o vanishes and only P^Q survives.

Since the kinetic and osmotic pressures are a measure of the kinetic and osmotic velocities dispersion, it is reasonable to assume an equation of state so that

$$P^{\mathrm{k}} + P^{\mathrm{o}} = P^{\mathrm{C}}(n), \qquad (4.25)$$

depending only on density. In this way, we obtain

$$P = P^{\mathrm{C}}(n) - \frac{\hbar^2 n}{4m} \frac{\partial^2}{\partial x^2} \ln n. \qquad (4.26)$$

For definiteness, we call P^{C} the "classical" part of the pressure, in the sense that it represents a measure of the velocities dispersion. However, it explicitly contains Planck's constant since it depends on P^{o} and the osmotic velocities, which have a purely quantum nature.

The rather drastic replacement of the sum of the kinetic and osmotic pressures by a function of the density only requires some comments. Equation (4.16) is exact but offer no advancement over the Wigner–Poisson formulation because ultimately it requires the knowledge of the ensemble wavefunctions. These, in turn, require the solution of a countable set of self-consistent Schrödinger equations, a hardly feasible task. In classical kinetic theory (therefore, without the osmotic and quantum pressures), it is customary to assume a closure assuming that the standard deviation of the velocities is a function of density only. The present suggestion just goes one step further, extending the usual approach to include the standard deviation of the osmotic velocities too. In addition, in the classical limit, we expect equations reproducing the classical fluid equations. This is certainly true if $P^{\mathrm{k}} + P^{\mathrm{o}} = P^{\mathrm{C}}(n)$. In other words, we expect the standard Euler equations to be reproduced thanks to the residual classical limit in P^{k}.

Notice that unlike P in (4.3) which uses the Wigner function, in (4.23) and (4.24) the statistical weights are provided by the $\tilde{p}_\alpha$ in (4.20). For a pure state one has $P^{\mathrm{C}} = 0$, which is in line with the understanding that a pure state corresponds to a cold plasma with no dispersion of velocities. The contribution P^{Q}, on the other hand, is a purely quantum pressure, with no classical counterpart.

It is well known that the closure problem is a delicate one. The derivation of macroscopic models from microscopic models always deserve some degree of approximation and a more or less phenomenological point of view. The present approach is capable of taking into account the quantum statistics of the charge carriers, represented by an appropriated equation of state. Moreover, it takes into account quantum diffraction effects, in particular tunneling and wave packet dispersion, present in the quantum part of the pressure. It is also able to reproduce the linear dispersion relation from kinetic theory with the exception of purely kinetic phenomena like Landau damping. Finally, the quantum fluid model reduces to the standard Euler equations in the formal classical limit and are sufficiently simple to be amenable to efficient numerical simulation.

Besides the above arguments, the proposed simplification is even more justified for an important class of statistical ensembles, where the wavefunctions have all equal (but not necessarily constant) amplitude,

$$\psi_\alpha = \sqrt{\frac{n}{N}}\, e^{iS_\alpha/\hbar}. \tag{4.27}$$

In this case, the osmotic pressure identically vanishes. Moreover, one has $\tilde{p}_\alpha = p_\alpha$, so that (4.23) becomes the usual standard deviation of the kinetic velocities. Hence, P^k can be interpreted in full analogy with the standard thermodynamic pressure. The velocities dispersion arises just from the randomness of the phases of the wavefunctions. The approximation can be viewed as a first step beyond the standard homogeneous equilibrium of a fermion gas, for which each state can be represented by a plane wave

$$\psi_\alpha(x,t) = A_0 \exp\left(\frac{imu_\alpha x}{\hbar}\right),$$

with the amplitude A_0 and the velocity u_α spatially constant. In the generalization (4.27), both the amplitude and the velocity can be spatially modulated, although we still restrict ourselves to the case where the amplitude is the same for all states. This appears to be a reasonable closure assumption for systems that are not too far from equilibrium.

In conclusion, with these hypothesis (4.25) the force equation (4.5) can be written as

$$\frac{\partial u}{\partial t} + u\frac{\partial u}{\partial x} = -\frac{1}{mn}\frac{\partial P^C(n)}{\partial x} + \frac{e}{m}\frac{\partial \phi}{\partial x} - \frac{1}{mn}\frac{\partial P^Q}{\partial x}. \tag{4.28}$$

Using the identity

$$\frac{1}{mn}\frac{\partial P^Q}{\partial x} = -\frac{\hbar^2}{2m^2}\frac{\partial}{\partial x}\left(\frac{\partial^2(\sqrt{n})/\partial x^2}{\sqrt{n}}\right), \tag{4.29}$$

we can rewrite the basic quantum hydrodynamic model for plasmas as composed by the continuity equation

$$\frac{\partial n}{\partial t} + \frac{\partial (nu)}{\partial x} = 0 \tag{4.30}$$

and the force equation

$$\frac{\partial u}{\partial t} + u\frac{\partial u}{\partial x} = -\frac{1}{mn}\frac{\partial P^C(n)}{\partial x} + \frac{e}{m}\frac{\partial \phi}{\partial x} + \frac{\hbar^2}{2m^2}\frac{\partial}{\partial x}\left(\frac{\partial^2(\sqrt{n})/\partial x^2}{\sqrt{n}}\right). \tag{4.31}$$

In the limit $\hbar \to 0$, this is formally equal to Euler's equation for an electron fluid in the presence of an electric field $-\partial \phi/\partial x$. Finally, we have the Poisson equation

$$\frac{\partial^2 \phi}{\partial x^2} = \frac{e}{\varepsilon_0}(n - n_0), \tag{4.32}$$

In comparison to the classical hydrodynamical equations for electrostatic plasmas, the difference is in the addition of the $\sim\hbar^2$ term in (4.31), the so-called Bohm potential term. While mathematically, the Bohm potential is equivalent to a pressure to be inserted in the momentum transport equation, physically it corresponds to typical quantum phenomena like tunneling and wave packet spreading. Therefore, it is not a pressure in the thermodynamic (velocities dispersion) sense.

Quantum hydrodynamical models have been derived in the context of semiconductor physics. For instance, Gardner [10] considered a quantum corrected displaced Maxwellian as introduced by Wigner [25]. More exactly, it is possible to derive the leading quantum correction $f_1(x,v,t)$ for a momentum-shifted local Maxwell–Boltzmann equilibrium $f_0(x,v,t)$, setting $f(x,v,t) = f_0(x,v,t) + \hbar^2 f_1(x,v,t)$ in the semiclassical quantum Vlasov equation (2.52) and collecting equal powers of $\hbar^2$. Here

$$f_0(x,v,t) = n(x,t) \left(\frac{m}{2\pi \kappa_B T(x,t)} \right)^{1/2} \exp \left(\frac{-m[v - u(x,t)]^2}{2\kappa_B T(x,t)} \right), \qquad (4.33)$$

on the assumption of a nondegenerate quasi-equilibrium state. Inserting the resulting Wigner function in the pressure defined in (4.3), a quantum hydrodynamical model similar to (4.30)–(4.32) is found, including also an energy transport equation for the temperature $T(x,t)$. By definition, the resulting system is restricted to quasi-Maxwellian, dilute systems. We also observe that in the case of a self-consistent problem, the electrostatic potential should have been also expanded in powers of some nondimensional quantum parameter. In conclusion, both procedures, involving a Madelung decomposition of the quantum ensemble wavefunctions or a quantum corrected Wigner function equilibrium, involve working hypotheses which are not rigorously justified. Starting from the Wigner–Poisson system, which is by definition collisionless, hardly one could derive rigorous macroscopic theories relying on quasi-equilibrium assumptions. Nevertheless, the numerical and analytical advantages of quantum fluid models over quantum kinetic models justify the popularity of macroscopic theories.

The quantum correction to the fluid equations, corresponding to the Bohm potential, was also derived from general thermodynamic arguments by Ancona and Tiersten in [3]. This work argues that the internal energy of the electron fluid in an electron–hole semiconductor should depend not only on the density but also on the density gradient, to extend the standard drift-diffusion model so as to include the quantum-mechanical behavior exhibited in strong inversion layers. Their theory defines a "double-force" and a "double-pressure vector" which allows for changes of the internal energy of the electron gas due purely to density fluctuations. Postulating a linear dependence of the double-pressure vector on density gradients (see (3.3) of [3] for more details) and working out the conservation laws of charge, mass, linear momentum and energy, Ancona and Tiersten found a generalized chemical potential composed of two contributions: (a) a gradient-independent term which can be modeled by the equation of state of a zero-temperature Fermi gas or any other appropriated form. This corresponds exactly to the classical pressure $P^C(n)$ of the quantum hydrodynamical model for plasmas; (b) a Bohm potential term proportional to a phenomenological parameter.

Later, Ancona and Iafrate [2] obtained the expression of the phenomenological coefficient of [3] from the Wigner formalism following the method of [10]. In other words, the first-order quantum correction for a Maxwell–Boltzmann equilibrium found from the semiclassical quantum Vlasov equation was employed to calculate the particle density and the stress tensor. Eliminating the potential function between the two expressions, the equation of state relating the stress tensor and the particle density and gradient is derived, containing the Bohm potential. The gradient-dependent term reflects the quantum mechanical nonlocality, with the equation of state depending on the derivatives of the density also. Once again, the demonstration in [2] is valid when quantum contributions to the self-consistent mean field potential can be ignored. Moreover, the dilute and semiclassical situation is supposed, where the Boltzmann statistics apply and a small dimensionless quantum parameter exist.

Further insight can be gained defining the enthalpy [11] or effective potential

$$W(n) = \int^n \frac{\mathrm{d}n'}{n'} \frac{\mathrm{d}P^C(n')}{\mathrm{d}n'}. \tag{4.34}$$

It is then possible to combine (4.30) and (4.31) into an effective single-particle Schrödinger equation. Indeed, let us define the effective wavefunction

$$\Psi(x,t) = \sqrt{n(x,t)} \exp\left(\mathrm{i}S(x,t)/\hbar\right), \tag{4.35}$$

normalized according to

$$\int |\Psi|^2 \mathrm{d}x = N. \tag{4.36}$$

The normalization is a matter of taste, but reflect the fact that $\Psi(x,t)$ is representative of the whole system even if it depends only on a single spatial coordinate x. Moreover the phase, or eikonal $S(x,t)$ is defined so that

$$u(x,t) = \frac{1}{m} \frac{\partial S(x,t)}{\partial x}. \tag{4.37}$$

We obtain that $\Psi(x,t)$ satisfies the equation

$$\mathrm{i}\hbar \frac{\partial \Psi}{\partial t} = -\frac{\hbar^2}{2m} \frac{\partial^2 \Psi}{\partial x^2} - e\phi\Psi + W\Psi. \tag{4.38}$$

This is a nonlinear Schrödinger equation, as the effective potential W depends on the wavefunction through (4.34), where $n = |\Psi|^2$. Separating (4.38) into its real and imaginary parts, we recover the continuity (4.30) and force (4.31) equations. Finally, the complete effective Schrödinger–Poisson system is composed by (4.38) and the Poisson equation

$$\frac{\partial^2 \phi}{\partial x^2} = \frac{e}{\varepsilon_0} \left(|\Psi|^2 - n_0 \right).$$ (4.39)

We notice that, in general, the dynamics of a statistical mixture must be treated with the full Wigner–Poisson system, or, equivalently, with a set of Schrödinger equations, coupled by Poisson's equation. We have shown that one can reduce the quantum transport problem to a single nonlinear Schrödinger equation plus Poisson's equation. The main result is that we can reduce the (phase space) Wigner problem to a nonlinear Schrödinger equation problem.

At this point, it is useful to remember (trying to be not excessively pedantic) the obvious limitations of the quantum hydrodynamic model given by (4.30)–(4.32): (a) it is a fluid model, and hence applicable only for long wavelengths. Kinetic phenomena requiring a detailed knowledge of the equilibrium Wigner function, such as Landau damping or recurrences in phase space (the plasma echo [19]) deserve a kinetic treatment; (b) no energy transport equation is included. However it would be a simple exercise to calculate it taking the second-order moment of the quantum Vlasov equation; (c) only nonrelativistic phenomena are included, since the starting point is the one-body Schrödinger equation with a mean field potential; (d) no spin effects are included, except for the equation of state which can, in a certain measure, represent some quantum statistical effects. This is the case if the equation of state for a dense, degenerate electron gas is assumed. In the same way, some relativistic effects can also be present if the equation of state is adapted to a relativistic electron gas; (e) no magnetic fields are allowable in the present formulation; (f) the sum of kinetic and osmotic pressures are replaced by an appropriated equation of state. For linear regimes, in the vicinity of homogeneous equilibria, this assumption works well, provided a reasonable equation of state is chosen.

4.3   Applications to Degenerate Plasma

Consider a zero-temperature one-dimensional electron gas, with Fermi velocity v_F and equilibrium density n_0. In this case, the classical pressure is

$$P^C = \frac{m v_F^2}{3 n_0^2} n^3.$$ (4.40)

Notice that the term "classical" is somewhat inappropriate here, as P^C will contain Planck's constant through the Fermi velocity. We postpone the derivation of convenient equations of state for a zero-temperature Fermi gas in one, two, and three spatial dimensions to Sect. 4.4. Additionally, in Sect. 4.4, we review some basic properties of Fermi gases in general.

Notice that in one spatial dimension, the Fermi velocity

$$v_F = \frac{\pi}{2} \frac{\hbar n_0}{m} \qquad (4.41)$$

is proportional to n_0, whereas in three dimensions $v_F \propto n_0^{1/3}$.
Using (4.40), the effective potential defined in (4.34) turns out to be

$$W = \frac{m v_F^2}{2 n_0^2} |\Psi|^4, \qquad (4.42)$$

taking into account $n = |\Psi|^2$.

The effective potential is repulsive, and tends to flatten the electron density. This is quite natural, as W derives from the pressure P^C, which in turn is a manifestation of the dispersion of velocities in a fermion gas.

We point out that a similar nonlinear Schrödinger equation with a $|\Psi|^4$-dependent potential has also been derived in the study of low-dimensional Bose condensates [16]. We stress, however, that such a boson-fermion duality only applies to one-dimensional systems.

4.3.1 Linear Wave Propagation

As a first application, let us study linear wave propagation for the quantum hydrodynamical model (4.30)–(4.32) with the pressure P^C as in (4.40). Linearizing around the homogeneous equilibrium

$$n = n_0, \quad u = 0, \quad e\phi = m v_F^2/2, \qquad (4.43)$$

we obtain the following dispersion relation (for waves with frequency ω and wavenumber k),

$$\omega^2 = \omega_p^2 + k^2 v_F^2 + \frac{\hbar^2 k^4}{4 m^2}. \qquad (4.44)$$

For $v_F = 0$, we recover the dispersion relation of the standard Schrödinger–Poisson system [6, 23]. Equation (4.44) can be written in dimensionless units,

$$\frac{\omega^2}{\omega_p^2} = 1 + k^2 \lambda_F^2 + \frac{k^4 \lambda_F^4}{4} \Gamma_Q. \qquad (4.45)$$

Here,

$$\lambda_F = \frac{v_F}{\omega_p}, \quad \Gamma_Q = \frac{\hbar^2 \omega_p^2}{m^2 v_F^4} \qquad (4.46)$$

are the Fermi length and quantum coupling parameter, adapted to one spatial dimension. Note that quantum mechanical effects (dispersion of the wave packet) are first-order in the coupling parameter Γ_Q, whereas quantum statistical effects (Fermi–Dirac distribution) appear at leading order.

We want to compare this dispersion relation (2.85) to the one obtained from the complete Wigner–Poisson system,

$$1 - \frac{\omega_p^2}{n_0} \int \frac{f_0(v)\,dv}{(\omega - kv)^2 - \hbar^2 k^4/4m^2} = 0. \tag{4.47}$$

In our case, $f_0(v)$ is given by the Fermi–Dirac distribution for a zero-temperature one-dimensional electron gas at equilibrium, that is, $f_0(v) = n_0/2v_F$ if $|v| < v_F$ and $f_0(v) = 0$ if $|v| > v_F$. Substituting into (4.47), one obtains (without any further approximation)

$$\frac{\omega^2}{\omega_p^2} = \frac{\Omega^2}{\omega_p^2} \coth\left(\frac{\Omega^2}{\omega_p^2}\right) + k^2 \lambda_F^2 + \frac{k^4 \lambda_F^4}{4}\, \Gamma_Q, \tag{4.48}$$

where

$$\frac{\Omega^2}{\omega_p^2} = \frac{\hbar k^3 v_F}{m \omega_p^2} = k^3 \lambda_F^3\, \Gamma_Q^{1/2}. \tag{4.49}$$

Now, we expand the first term on the right-hand side of (4.48) in the long wavelength (fluid) limit $\Omega \ll \omega_p$. Using the expansion $x \coth(x) = 1 + x^2/3 - x^4/45 + \cdots$, one obtains

$$\frac{\omega^2}{\omega_p^2} = 1 + k^2 \lambda_F^2 + \left(\frac{k^4 \lambda_F^4}{4} + \frac{k^6 \lambda_F^6}{3}\right) \Gamma_Q - \frac{1}{45} k^{12} \lambda_F^{12}\, \Gamma_Q^2 + \cdots. \tag{4.50}$$

This is a double expansion in powers of the parameters Γ_Q and $k\lambda_F$. The collisionless regime is in principle characterized by $\Gamma_Q \ll 1$, although, as was seen in Sect. 2.5, electron–electron interactions can be neglected even when $\Gamma_Q \simeq 1$, as is the case for metals. On the other hand, the fluid regime is characterized by small wave numbers ($\Omega \ll \omega_p$). Indeed, keeping terms to fourth-order in $k\lambda_F$, (4.50) reduces to the dispersion relation for the effective Schrödinger–Poisson system, (4.45). This is a further indication that the effective Schrödinger–Poisson system is a good approximation to the complete Wigner–Poisson system for long wavelengths.

We also note that for $\Gamma_Q \to 0$, the dispersion relation reduces to

$$\omega^2 = \omega_p^2 + k^2 v_F^2. \tag{4.51}$$

This is exactly the dispersion relation obtained from the classical Vlasov–Poisson system with a zero-temperature Fermi–Dirac equilibrium. In other words, when the quantum coupling parameter is vanishingly small, a classical dynamical equation can be used, as the only quantum effects come from the Fermi–Dirac statistics.

4.3.2 Stationary Solutions

As a second illustration, we use the present formalism to describe the stationary states of the electron gas [13]. This result is more easily obtained by using the fluid version of our model. In the time-independent case, the continuity equation (4.30) and the force equation (4.31) possess the following first integrals

$$J = A^2 u, \quad \Xi = \frac{mu^2}{2} - e\phi + W - \frac{\hbar^2}{2mA}\frac{d^2A}{dx^2}, \tag{4.52}$$

where $A = \sqrt{n} = |\Psi|$. The first integrals in (4.52) corresponds to current (J) and energy (Ξ) conservation. We can always choose $\Xi = 0$ by a shift in the electrostatic potential. In this way, we can reduce the description of the stationary states to a set of second-order nonlinear ordinary differential equations for the amplitude A and the electrostatic potential ϕ. For a zero-temperature one-dimensional electron gas, the effective potential W is given by (4.42), or

$$W = \frac{mv_F^2 n^2}{2n_0^2}. \tag{4.53}$$

Hence, from (4.39) and (4.52), we get

$$\hbar^2 \frac{d^2A}{dx^2} = m\left(\frac{mJ^2}{A^3} - 2eA\phi + \frac{mv_F^2}{n_0^2}A^5\right), \tag{4.54}$$

$$\frac{d^2\phi}{dx^2} = \frac{e}{\varepsilon_0}(A^2 - n_0). \tag{4.55}$$

Notice that, if the amplitude $A(x)$ is a slowly varying function of x, the second derivative on the left-hand side of (4.54) can be neglected. With this assumption, (4.54) reduces to an algebraic equation, which can be solved for A, and the result plugged into (4.55). This becomes a nonlinear differential equation for the electrostatic potential, which is the Thomas–Fermi approximation [4] to the fluid model.

Let us compare (4.54) and (4.55) with (3.22) and (3.23) derived in the context of the quantum Dawson model for a one-stream plasma. Now in (4.54) there is the presence of a Fermi pressure term which contribute to the flattening of the amplitude. This reflect that the quantum Dawson model is intrinsically a "cold" model, without the possibility of including any classical pressure contribution. Moreover there is a difference of philosophy: we are not dealing with a "one-stream" plasma but with a full quantum mixture which, at the end, can all be reduced to an effective wavefunction $\Psi(x,t)$.

It can be verified that the $J = 0$ case cannot sustain small-amplitude, periodic solutions. Hence, we assume $J = n_0 u_0$ with $u_0 \neq 0$ and introduce the following rescaling, similar to (3.24) but now with a rescaled Fermi velocity also,

$$\hat{x} = \frac{\omega_p x}{u_0}, \quad \hat{A} = \frac{A}{\sqrt{n_0}}, \quad \hat{\phi} = \frac{e\phi}{m u_0^2}$$

$$H = \frac{\hbar \omega_p}{m u_0^2}, \quad V_F = \frac{v_F}{u_0}. \tag{4.56}$$

We obtain, in the transformed variables (omitting the circumflex for simplicity of notation),

$$H^2 \frac{d^2 A}{dx^2} = -2\phi A + \frac{1}{A^3} + V_F^2 A^5, \tag{4.57}$$

$$\frac{d^2 \phi}{dx^2} = A^2 - 1, \tag{4.58}$$

a system that depends only on the rescaled parameters H and V_F. Note that the quantum coupling parameter can be written as $\Gamma_Q = H/V_F^2$.

It is interesting to perform a linear stability analysis to see in what conditions the system supports small amplitude spatially periodic solutions. Writing

$$A = 1 + A' \exp(ikx), \quad \phi = \frac{(1 + V_F^2)}{2} + \phi' \exp(ikx), \tag{4.59}$$

and retaining in (4.57) and (4.58) only terms up to first-order in the primed variables, we obtain the relation

$$H^2 k^4 - 4(1 - V_F^2)k^2 + 4 = 0. \tag{4.60}$$

This second degree equation has solutions

$$k^2 = \frac{2(1 - V_F^2) \pm 2\sqrt{(1 - V_F^2)^2 - H^2}}{H^2}. \tag{4.61}$$

Clearly, spatially oscillating solutions only exist when k^2 is real and positive, which yields the condition

$$V_F^2 < 1 - H, \tag{4.62}$$

or equivalently

$$m u_0^2 > m v_F^2 + \hbar \omega_p. \tag{4.63}$$

This expression sets a lower bound on the speed u_0, below which no oscillating stationary solution can exist. Hence, the Fermi pressure tends to suppress these linear oscillations, as is natural since the exclusion principle tends to flatten the probability distribution (to compare with the $H < 1$ condition from the one-stream quantum Dawson model).

4.3.3   Two-Stream Instability

A classical plasma composed of two counterstreaming electronic populations with velocities $\pm u_0$ can give rise, for certain wavenumbers, to an instability. In the previous chapter, we have shown that quantum effects modify the dispersion relation, and give rise to a new instability branch. These results were obtained by neglecting the effects of quantum statistics, and are therefore valid in the limit $v_F \ll u_0$. Here, we perform the same calculations for finite values of v_F.

We consider two electronic populations, which are both distributed according to a zero-temperature Fermi–Dirac equilibrium, but with average velocities $\pm u_0$. The motionless ions provide a neutralizing background. The dispersion relation for such a two-stream plasma can be found in the following way. For a single stream propagating at velocity $\pm u_0$, our fluid model yields the following dielectric constant (thus valid for long wavelengths)

$$\varepsilon_{\pm}(k,\omega) = 1 - \frac{\omega_p^2}{(\omega \mp k u_0)^2 - k^2 v_F^2 - \hbar^2 k^4/4m^2}. \tag{4.64}$$

Setting $\varepsilon_{\pm}(k,\omega) = 0$ leads to the dispersion relation found previously, (4.44), with the appropriate Doppler shift. The dielectric constant for the two-stream case is found by averaging the contributions from each stream $\varepsilon(k,\omega) = (\varepsilon_+ + \varepsilon_-)/2$. Using the normalization of (4.56), we obtain

$$\varepsilon(k,\omega) = 1 - \frac{1/2}{(\omega+k)^2 - k^2 V_F^2 - H^2 k^4/4} - \frac{1/2}{(\omega-k)^2 - k^2 V_F^2 - H^2 k^4/4}. \tag{4.65}$$

Setting $\varepsilon(k,\omega) = 0$, we obtain the dispersion relation for the two-stream plasma

$$\omega^4 - \left(1 + 2k^2(1+V_F^2) + \frac{H^2 k^4}{2}\right)\omega^2$$

$$- k^2\left(1 - V_F^2 - \frac{H^2 k^2}{4}\right)\left(1 - (1-V_F^2)k^2 + \frac{H^2 k^4}{4}\right) = 0. \tag{4.66}$$

Notice that for $V_F = 0$ we recover the dispersion relation obtained in [13]. Solving for ω^2, one obtains

$$\omega^2 = \frac{1}{2} + k^2\left(1 + v_F^2 + \frac{H^2 k^2}{4}\right) \pm \frac{1}{2}\left[1 + 8k^2\left(1 + 2k^2 V_F^2 + \frac{H^2 k^4}{2}\right)\right]^{1/2}. \tag{4.67}$$

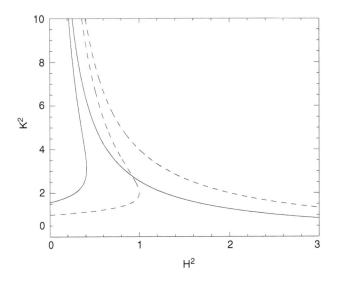

Fig. 4.1 Stability diagram for the two-stream plasma, with $V_F = 0.7$ (*solid lines*) and $V_F = 0$ (*dashed lines*) according to [20]. The curves correspond to (4.69) and (4.70). For both cases, the region of the plane containing the H^2 axis is unstable

The solution for ω^2 has two branches, one of which is always positive and gives stable oscillations. The other solution is negative ($\omega^2 < 0$) provided that

$$[H^2 k^2 - 4(1 - V_F^2)][H^2 k^4 - 4(1 - V_F^2)k^2 + 4] < 0. \qquad (4.68)$$

We immediately notice that, if $V_F \geq 1$, (4.68) is never verified, and therefore there is no instability. This is a quite natural result. Indeed, mathematically, the instability is due to the fact that the two-stream velocity distribution has a "hole" around $v = 0$. When $V_F \geq 1$, the hole is filled up, and no instability can occur. To put it differently, there can be instability only when the equilibrium distribution is a nonmonotonic function of the energy, which ceases to be true when $V_F \geq 1$.

When $V_F < 1$, (4.68) bifurcates for $H = 1 - V_F^2$. If $H \geq 1 - V_F^2$, the second factor is always positive, and instability occurs for $H^2 k^2 < 4(1 - V_F^2)$. If $H < 1 - V_F^2$, there is instability if either

$$0 < H^2 k^2 < 2(1 - V_F^2) - 2\sqrt{(1 - V_F^2)^2 - H^2}, \qquad (4.69)$$

or

$$2(1 - V_F^2) + 2\sqrt{(1 - V_F^2)^2 - H^2} < H^2 k^2 < 4(1 - V_F^2). \qquad (4.70)$$

This yields the stability diagram plotted in Fig. 4.1, generalizing the result obtained [13] in the limiting case $V_F = 0$, see Fig. 4.1. The presence of a finite Fermi velocity has the effect of reducing the region of instability. Numerical simulations yield similar result to those observed in the $V_F = 0$ case, which are reported in [13].

4.4 Equation of State for a Zero-Temperature Fermi Gas

How to chose the equation of state is not always an evident subject. However, two necessary requirements are: (a) the linear dispersion relation from the fluid equations should reproduce the long wavelength (fluid) limit of linear dispersion relation for wave propagation in kinetic theory and (b) the function $P^C(n)$ should coincide with the pressure calculated from kinetic theory in equilibrium.

Therefore, a first task is to find the dispersion relation from kinetic theory. For completeness, the one, two, and three dimensional cases will be considered, since there are some small numerical differences according to the number of degrees of freedom. Focusing on electrostatic waves, we have seen that linear wave propagation is described in (2.85). For arbitrary dimensionality, it is

$$\varepsilon = 1 - \frac{\omega_p^2}{n_0} \int \frac{f_0(\mathbf{v})d\mathbf{v}}{(\omega - \mathbf{k} \cdot \mathbf{v})^2 - \hbar^2 k^4/(4m^2)} = 0. \tag{4.71}$$

For a zero-temperature, completely degenerate Fermi gas, the equilibrium Wigner function is

$$f_0(v) = \frac{n_0}{2v_F}, \quad |v| < v_F,$$

$$f_0(v) = 0, \quad |v| > v_F, \quad (1D), \tag{4.72}$$

in the one-dimensional case. Inserting this functional dependence on the dielectric constant interpreted in the principal value sense gives

$$\varepsilon = 1 - \frac{m\omega_p^2}{2\hbar k^3 v_F} \ln \left| \frac{\omega^2 - (kv_F - \hbar k^2/(2m))^2}{\omega^2 - (kv_F + \hbar k^2/(2m))^2} \right| = 0. \tag{4.73}$$

In the long wavelength limit $kv_F \ll \omega, \hbar k^2/(2m) \ll \omega$, expanding ε gives

$$\varepsilon = 1 - \frac{\omega_p^2}{\omega^2} \left(1 + \frac{k^2 v_F^2}{\omega^2} + \frac{\hbar^2 k^4}{4m^2 \omega^2} \right) = 0, \tag{4.74}$$

or

$$\omega^2 = \omega_p^2 + k^2 v_F^2 + \frac{\hbar^2 k^4}{4m^2}, \quad (1D) \tag{4.75}$$

which is the limit of the kinetic dispersion relation for small wavenumber and one spatial dimension.

For the analogous situation in two spatial dimensions, we consider (4.71) with the equilibrium distribution

$$f_0(\mathbf{v}) = \frac{n_0}{\pi v_F^2}, \quad |\mathbf{v}| < v_F,$$

$$f_0(\mathbf{v}) = 0, \quad |\mathbf{v}| > v_F, \quad (2D). \tag{4.76}$$

Proceeding as in the one-dimensional case, the long wavelength dispersion relation is found to be

$$\omega^2 = \omega_p^2 + \frac{3}{4}k^2 v_F^2 + \frac{\hbar^2 k^4}{4m^2}. \quad (2D) \tag{4.77}$$

Notice the different numerical factor in front of the Fermi velocity contribution.

To finalize, in three spatial dimensions we use

$$f_0(\mathbf{v}) = \frac{n_0}{4\pi v_F^3/3}, \quad |\mathbf{v}| < v_F,$$

$$f_0(\mathbf{v}) = 0, \quad |\mathbf{v}| > v_F, \quad (3D), \tag{4.78}$$

yielding

$$\omega^2 = \omega_p^2 + \frac{3}{5}k^2 v_F^2 + \frac{\hbar^2 k^4}{4m^2}. \quad (3D) \tag{4.79}$$

To sum up, the dispersion relation for a completely degenerate Fermi gas in the long wavelength limit is

$$\omega^2 = \omega_p^2 + \frac{3}{D+2}k^2 v_F^2 + \frac{\hbar^2 k^4}{4m^2}, \tag{4.80}$$

where D is the number of degrees of freedom. Whatever the equation of state in the fluid equations, it should be able to reproduce (4.80) after linearizing the hydrodynamic model.

It is convenient to write here the three-dimensional version of the quantum fluid model,

$$\frac{\partial n}{\partial t} + \nabla \cdot (n\mathbf{u}) = 0, \tag{4.81}$$

$$\frac{\partial \mathbf{u}}{\partial t} + \mathbf{u} \cdot \nabla \mathbf{u} = -\frac{1}{mn}\nabla \cdot \mathbf{P}^C + \frac{e}{m}\nabla \phi + \frac{\hbar^2}{2m^2}\nabla\left(\frac{\nabla^2(\sqrt{n})}{\sqrt{n}}\right), \tag{4.82}$$

$$\nabla^2 \phi = \frac{e}{\varepsilon_0}(n - n_0), \tag{4.83}$$

where $\mathbf{P}^C = \mathbf{P}^C(n)$ is the sum of kinetic and osmotic velocities as before. The derivation of the three-dimensional quantum hydrodynamic model follows after taking the first- and second-order moments of the quantum Vlasov equation in three spatial dimensions. In this way, a pressure dyad $\mathbf{P}^C$ instead of a pressure function will enter the momentum equation. After Madelung-decomposing the Wigner function, the pressure dyad can be shown to be composed by a kinetic pressure dyad, an osmotic pressure dyad, and a Bohm potential contribution. As in the one-dimensional case, we assume the sum of the kinetic and osmotic parts to be entirely determined by the equilibrium Wigner function. In this way, the classical limit of the model is assured to be given as in the classical case, where the pressure dyad – and hence the equation of state – are determined by the equilibrium distribution function.

To conclude, now we have a classical pressure dyad, with components P_{ij}^C determined by the equilibrium Wigner function according to

$$P_{ij}^C = m \int d\mathbf{v}\, f_0(\mathbf{v})(v_i - u_i)(v_j - u_j). \tag{4.84}$$

Here, the u_i are the components of the mean fluid flow velocity,

$$n\mathbf{u} = \int f_0(\mathbf{v})\mathbf{v}d\mathbf{v}, \tag{4.85}$$

while n is the fluid density or zeroth-order moment of the equilibrium Wigner function.

For isotropic equilibria, one get

$$P_{ij}^C = P^C \delta_{ij}, \tag{4.86}$$

where

$$P^C = P^C(n) = \frac{1}{D} Tr[\mathbf{P}^C] \tag{4.87}$$

is the scalar pressure, with Tr denoting the trace. For isotropic equilibria, the choice of the equation of state amounts to the choice of a suitable scalar pressure.

For a completely degenerate Fermi gas, from (4.84) and (4.87) the scalar pressure in equilibrium turns out to be

$$P^C = \frac{mn_0 v_F^2}{D+2}, \tag{4.88}$$

with a dimension-dependent numerical factor.

Now it comes the crucial point. The equation of state should be consistent with the linear wave propagation properties from (4.80), and with the equilibrium scalar pressure from (4.88). A reasonable choice is a polytropic equation of state

$$P^C(n) = \alpha mn_0 v_F^2 \left(\frac{n}{n_0}\right)^\beta, \tag{4.89}$$

where α and β are coefficients to be determined. The above polytropic form imply a dispersion relation

$$\omega^2 = \omega_p^2 + \alpha\beta k^2 v_F^2 + \frac{\hbar^2 k^4}{4m^2} \tag{4.90}$$

from the quantum fluid equations (4.81)–(4.83). Consistency with (4.80) and (4.88) is achieved if and only if

$$\alpha = \frac{1}{D+2}, \quad \beta = 3, \tag{4.91}$$

Fig. 4.2 Quasi-equilibrium
distribution function $f_0(x,v,t)$
for an one-dimensional
zero-temperature Fermi gas,
where $n = n(x,t), u = u(x,t)$
and $u_F = u_F(x,t)$ are resp. the
local number density, mean
velocity and Fermi velocity

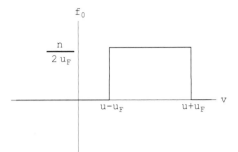

so that

$$P^C(n) = \frac{mn_0 v_F^2}{D+2}\left(\frac{n}{n_0}\right)^3,$$
(4.92)

is the desired equation of state. For $D = 1$, (4.40) is recovered.

Since no energy transport equation was included in the model, β can be interpreted as the adiabatic coefficient $c_p/c_v = (D+2)/D$, where c_p and c_v are the specific heats at constant pressure and constant volume, respectively. An adiabatic equation of state appears because the model apply to fast phenomena where there is no available time to heat transfer, also justifying the noninclusion of an energy transport equation. Hence, $c_p/c_v = 3$ correspond to an one-dimensional ($D = 1$) adiabatic coefficient, reflecting the fact that wave propagation is essentially an one-dimensional phenomena [21]. The treatment can be extended to nonzero temperatures using Fermi–Dirac integrals, but we keep $T = 0$ where the basic ideas are already illustrated.

Notice that any positive-definite, monotonous, differentiable function of the form

$$P^c(n) = \frac{mn_0 v_F^2}{D+2}\Pi\left(\frac{n}{n_0}\right)$$
(4.93)

would work equally well, provided the function $\Pi(n/n_0)$ satisfies

$$\Pi(1) = 1, \quad \Pi'(1) = 3.$$
(4.94)

Equivalently, the equation of state would be uniquely determined if and only if all derivatives of $\Pi(n/n_0)$ were known. This amounts to work out all of the higher-order moment equations, an unfeasible task.

The equation of state in (4.40) for an one-dimensional zero-temperature Fermi gas can also be justified as follows. For a zero-temperature system, we can assume a local quasi-equilibrium distribution function $f_0(x,v,t)$ as in Fig. 4.2. In other words, $f_0(x,v,t) = n(x,t)/[2u_F(x,t)]$ for $u(x,t) - u_F(x,t) \leq v \leq u(x,t) + u_F(x,t)$ and $f_0(x,v,t) = 0$ otherwise. Therefore, from construction, we identify the local density $n(x,t)$, the local velocity $u(x,t)$ and the local Fermi velocity $u_F(x,t)$.

We postulate that the classical pressure comes only from the equilibrium Wigner function,

$$P^C = m \left(\int f_0(x,v,t)v^2 dv - n(x,t)u(x,t)^2 \right), \tag{4.95}$$

yielding

$$P^C = \frac{mn(x,t)u_F(x,t)^2}{3}. \quad (1D) \tag{4.96}$$

The use of the equilibrium distribution function assures that the fluid equations reproduce the classical Euler equations when $\hbar \to 0$.

To reproduce (4.92) in the $D = 1$ case, we define the linear dependence

$$u_F(x,t) = \frac{v_F}{n_0}n(x,t) \Rightarrow P^C = \frac{mn_0 v_F^2}{3}\left(\frac{n}{n_0}\right)^3. \tag{4.97}$$

The reasoning for two and three spatial dimensions is similar. In two dimensions, postulate a quasi-equilibrium Wigner function $f_0(\mathbf{r},\mathbf{v},t) = n(\mathbf{r},t)/[\pi u_F(\mathbf{r},t)^2]$ in the circle $|\mathbf{v}-\mathbf{u}(\mathbf{r},t)|^2 \le u_F^2(\mathbf{r},t)$ and zero otherwise. The scalar pressure turns out to be

$$P^C = \frac{mn(\mathbf{r},t)u_F^2(\mathbf{r},t)}{4} = \frac{mn_0 v_F^2}{4}\left(\frac{n(\mathbf{r},t)}{n_0}\right)^3, \quad (2D) \tag{4.98}$$

provided the linear dependence in (4.97) is satisfied.

In three dimensions, we have $f_0(\mathbf{r},\mathbf{v},t) = 3n(\mathbf{r},t)/[4\pi u_F^3(\mathbf{r},t)]$ for $|\mathbf{v}-\mathbf{u}(\mathbf{r},t)|^2 \le u_F^2(\mathbf{r},t)$ and zero otherwise. We get

$$P^C = \frac{mn(\mathbf{r},t)u_F^2(\mathbf{r},t)}{5} = \frac{mn_0 v_F^2}{5}\left(\frac{n(\mathbf{r},t)}{n_0}\right)^3. \quad (3D) \tag{4.99}$$

once again from (4.97) chosen so as to comply with (4.92).

In passing, it is illustrative to calculate the Fermi velocity v_F according to the number of degrees of freedom. Suppose N fermions confined in a spatial length L, so that the equilibrium number density is $n_0 = N/L$. We need to distribute the N particles in a phase space of area $2mv_F L$, corresponding to $2mv_F L/h$ cells where h is Planck's constant. Since each cell can be populated by two fermions with opposite spin, we have

$$\frac{2mv_F L}{h} = \frac{N}{2} \Rightarrow v_F = \frac{\pi \hbar n_0}{2m}, \quad (1D) \tag{4.100}$$

showing a linear dependence between Fermi velocity and density in one spatial dimension.

Applying the same method, we found

$$v_F = \frac{\hbar}{m}(2\pi n_0)^{1/2}, \quad (2D) \tag{4.101}$$

$$v_F = \frac{\hbar}{m}(3\pi^2 n_0)^{1/3}, \quad (3D). \tag{4.102}$$

Notice the number density has different units according to the dimensionality D : $n_0 \sim 1/L^D$ where L is a length scale.

4.5   Landau Damping in a Degenerate Plasma

The neglecting of kinetic effects is a necessary assumption for the use of hydrodynamic equations. Even if the need for a kinetic treatment deserves to be analyzed
in each particular situation, we can consider the case of the propagation of high
frequency electrostatic waves in a degenerate plasma. In this way, we can measure
the importance of the Landau damping, a kinetic effect not taken into account in
usual fluid models. As we shall shortly see, quantum Langmuir waves are strongly
damped for wavenumbers smaller than the Fermi wavenumber. Hence, a fluid
treatment can be valid only for long wavelengths, larger than the Thomas–Fermi
length.

Generalizing (2.93) of Chap. 2 to three dimensions, we have the following
dispersion relation for electrostatic waves,

$$\varepsilon = 1 - \frac{m\,\omega_p^2}{n_0\,\hbar k^2}\left(\int_{L_+}\frac{\mathrm{d}\mathbf{v}\,f_0(\mathbf{v})}{k[v_z - \hbar k/(2m)] - \omega} - \int_{L_-}\frac{\mathrm{d}\mathbf{v}\,f_0(\mathbf{v})}{k[v_z + \hbar k/(2m)] - \omega}\right) = 0,$$
(4.103)

where a wave vector $\mathbf{k}k\hat{z}$ was assumed without loss of generality. Here, the
equilibrium distribution Wigner function is $f_0(\mathbf{v})$ and $L_\pm$ are Landau contours under
the poles $v_z = \omega/k \pm \hbar k/(2m)$.

Defining the projection

$$f_{0z}(v_z) = \int \mathrm{d}v_x \mathrm{d}v_y f_0(\mathbf{v})$$
(4.104)

we obtain,

$$\varepsilon = 1 - \frac{m\,\omega_p^2}{n_0\,\hbar k^2}\left(\int_{L_+}\frac{\mathrm{d}v_z f_0(v_z)}{k[v_z - \hbar k/(2m)] - \omega} - \int_{L_-}\frac{\mathrm{d}v_z f_0(v_z)}{k[v_z + \hbar k/(2m)] - \omega}\right) = 0,$$
(4.105)

which is formally the same as (2.93). Therefore, in analogy with (2.95), we have

$$\gamma \simeq \gamma_{\mathrm{cl}} = \frac{\pi\,\omega_p^3}{2n_0 k^2}\frac{\mathrm{d}f_{0z}}{\mathrm{d}v_z}\left(v_z = \frac{\omega}{k}\right).$$
(4.106)

as the classical limit of the growth rate γ. The more exact growth rate expression
(2.94) could have been employed, but quantum diffraction effects are not so
fundamental in the present discussion.

In conclusion, from (4.106) we see that damping (or growth) is determined by the
symbol of the derivative of the projected equilibrium Wigner function. To employ
this result to a degenerate plasma, it is not so useful to focus on a zero-temperature
plasma because in this case the equilibrium distribution is not differentiable. Rather,

we consider the Thomas–Fermi expression for the equilibrium Wigner distribution function [9, 14],

$$f_0(\mathbf{v}) = \frac{\alpha}{\exp\left[\beta\left(\frac{mv^2}{2} - \mu\right)\right] + 1}, \tag{4.107}$$

where $v^2 = v_x^2 + v_y^2 + v_z^2$, $\beta = 1/(\kappa_B T)$ and the normalization constant is

$$\alpha = -\frac{n_0}{Li_{3/2}(-e^{\beta\mu})}\left(\frac{m\beta}{2\pi}\right)^{3/2} = 2\left(\frac{m}{2\pi\hbar}\right)^3. \tag{4.108}$$

Here μ is the chemical potential, which is found solving the last equality in (4.108). In the limit of a vanishing temperature T, one has $\mu \simeq E_F = mv_F^2/2$, the Fermi energy of the system. Moreover, $Li_{3/2}$ is a polylogarithm function of argument $3/2$. Equation (4.108) is found from

$$n_0 = \int d\mathbf{v}\, f_0(\mathbf{v}) = 4\pi\alpha\left(\frac{2}{\beta m}\right)^{3/2}\int_0^\infty \frac{u^2\, du}{\exp[-\beta\mu]u^2 + 1}$$

$$= -\alpha\left(\frac{2\pi}{\beta m}\right)^{3/2}Li_{3/2}(-e^{\beta\mu}). \tag{4.109}$$

Performing the integration over the perpendicular velocity components, we get

$$f_{0z}(v_z) = \frac{2\pi\alpha}{m\beta}\ln\left(1 + \exp\left[\beta\left(\mu - \frac{mv_z^2}{2}\right)\right]\right), \tag{4.110}$$

which has a bell-shaped profile as can be verified. For the corresponding damping rate, using (4.106), we obtain

$$\frac{\gamma}{\omega_p} = -\frac{\pi^2\alpha\omega_p^3}{n_0 k^3}\left(1 + \exp\left[\beta\left(\frac{m\omega_p^2}{2k^2} - \mu\right)\right]\right)^{-1}, \tag{4.111}$$

where the replacement $\omega \simeq \omega_p$ was employed.

For very small temperature, we have $\mu \simeq E_F$ and a very large β. From (4.111) it follows that for $\omega_p^2/k^2 > 2E_F/m$ or equivalently for $\omega_p/k > v_F$ there will be no damping at all. This is because the phase velocity of the wave is in a region where there are no particles to interact.

In the opposite case when $\omega_p/k < v_F$ the exponential in (4.111) will be zero for very large β, producing significant damping not taken into account in the fluid description. However, for $\omega_p/k < v_F$ one has, using the definition (4.108) of α and the expression (4.102) of the Fermi velocity,

$$\frac{|\gamma|}{\omega_p} = \frac{1}{4\pi n_0}\left(\frac{m\omega_p}{\hbar k}\right)^3 < \frac{1}{4\pi n_0}\left(\frac{mv_F}{\hbar}\right)^3 = \frac{3\pi}{4}, \tag{4.112}$$

which is not particularly of any help. Therefore, to avoid damping, for quantum Langmuir waves in degenerate plasma necessarily the long wavelength assumption

$$k < \frac{\omega_p}{v_F} \tag{4.113}$$

must hold, a condition usually not remembered in the literature. Otherwise, a kinetic modeling is necessary. The reason for the damping is that many particles would be available to interact with the wave if (4.113) is not satisfied. Notice that this result does not hold in the case of an one-dimensional degenerate Fermi gas, where the derivative of the equilibrium Wigner function is almost everywhere zero. The above reasoning is equally valid in the case of nondegenerate plasma, provided the Fermi velocity is replaced by the thermal velocity.

4.6   Decomposing an Equilibrium Wigner Function in Terms of Ensemble Wavefunctions

The determination of the equilibrium reduced one-body Wigner function associated with an equilibrium density matrix follows from the Wigner transform as defined in Chap. 2. For instance [7], in the case of equilibrium with a heat bath at temperature T one has the equilibrium density operator $\hat{\rho}$ given by

$$\hat{\rho} = \frac{e^{-\beta \hat{H}}}{Tr[e^{-\beta \hat{H}}]}, \tag{4.114}$$

where $\hat{H}$ is the Hamiltonian operator, $\beta = 1/(\kappa_B T)$ and Tr denotes the trace. Correspondingly, the density matrix is

$$\rho(x, y) = \sum_n p_n \varphi_n^*(x) \varphi_n(y), \tag{4.115}$$

where the $\varphi_n(x)$ are the normalized eigenfunctions of the Hamiltonian operator with eigenvalues E_n, or

$$\hat{H} |\varphi_n> = E_n |\varphi_n> . \tag{4.116}$$

In (4.115),

$$p_n = \frac{e^{-\beta E_n}}{Z} \tag{4.117}$$

are the statistical weights for the canonical ensemble. Here, Z is the canonical partition function,

$$Z = \sum_n e^{-\beta E_n}. \tag{4.118}$$

The Hamiltonian operator is frequently composed by a kinetic energy term and some potential $V(x)$ which typically is the sum of a mean field, self-consistent term, and an external potential. We are using the index n instead of α to label the ensemble quantum states to be more in accordance with the usual quantum mechanics notation in the present context. Also, we are omitting the time t in all equations of this section since only stationary problems are discussed.

In self-consistent problems, the task of finding the equilibrium eigenfunctions is already a difficult problem which can be done only numerically, with some optimism. However, ignoring the self-consistent potential, with some luck the density matrix (4.115) can be explicitly written, for sufficiently simple external potentials. Then, the reduced one-body Wigner function $f_0(x,v)$ can be found through (2.8)

$$f_0(x,v) = \frac{Nm}{2\pi\hbar} \int ds\, e^{\frac{imvs}{\hbar}} \rho\left(x+\frac{s}{2}, x-\frac{s}{2}\right), \qquad (4.119)$$

adapted to the normalization $\int dxdv f_0(x,v) = N$.

The inverse problem of determining the quantum statistical ensemble from a given equilibrium reduced one-body Wigner function, however, is more appealing from the plasma physics point of view. Indeed, traditionally in plasma one starts from a given equilibrium distribution function and then proceed to the study of waves, instabilities and so on. For definiteness, consider an spatially homogeneous equilibrium $f_0(v)$. Equation (2.7) then gives the density matrix

$$\rho(x,y) = \frac{1}{N} \int dv\, e^{\frac{imv(x-y)}{\hbar}} f_0(v). \qquad (4.120)$$

Equation (4.120) suggest a recipe to calculate the statistical weights in the quantum ensemble corresponding to $f_0(v)$, once the quantum states are chosen. For example, consider the case of a momentum-shifted one-dimensional zero-temperature Fermi gas equilibrium where $f_0(v) = n_0/[2v_F]$ for $u - v_F \leq v \leq u + v_F$ and $f_0(v) = 0$ otherwise. Here, the density u_0, the mean velocity u and the Fermi velocity v_F will be assumed constant. Using (4.120), we get

$$\rho(x,y) = \frac{n_0\hbar}{Nmv_F(x-y)} \sin\left(\frac{mv_F(x-y)}{\hbar}\right) \exp\left(\frac{imu(x-y)}{\hbar}\right), \qquad (4.121)$$

displaying an oscillatory pattern. For comparison, see Fig. 4.2 showing the Wigner function and Fig. 4.3 showing N/n_0 times the real part of the density matrix for $u/v_F = 3$. Notice that (4.121) does not satisfy the condition (2.69) for a pure state, except in the limit $v_F \to 0$. By the way, it is more proper to refer to $v_F \to 0$ as the dilute rather than the cold limit, since v_F is a function of the density and not of the thermodynamic temperature, which is assumed to be identically null in the example.

To represent the microscopic structure of the density matrix in terms of quantum states and corresponding occupation probabilities, a definite set of wavefunctions

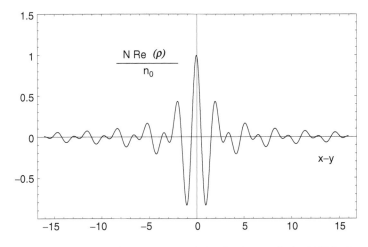

Fig. 4.3 Oscillatory structure of the real part of the density matrix in (4.121). Lengths are measured in units of the Fermi length $L_F = \hbar/(mv_F)$. In the graphic, $u/v_F = 3$. A similar pattern appear for the imaginary part

need to be selected. For unbounded plasma, though not mandatory, it is natural to choose the plane waves

$$\psi_k(x) = A\exp(ikx), \quad -\infty < k < \infty \tag{4.122}$$

where the amplitude $A = 1/\sqrt{L}$. Here, L is the fictitious length of the system, eventually set to infinity at the end of the calculation.

In the case of confined charged particle systems, as in semiconductor quantum wells or other nanoscopic systems, a different set of quantum states can be a better choice depending on the confinement. For instance, in the case of parabolic quantum wells with harmonic external potential, the harmonic oscillator eigenfunctions play a distinguished rôle.

So, we have the plane waves (4.122), each with some occupation probability $p(k) \geq 0$. Since here k is a continuous label, we now need $\int dk\, p(k) = 1$. Adapting (2.7) to the continuous label case, we have

$$\rho(x,y) = \int dk\, p(k)\psi_k(x)\,\psi_k^*(y) = A^2 \int dk\, p(k)\exp(ik(x-y)). \tag{4.123}$$

The equality of (4.121) and (4.123) imply

$$p(k) = \frac{\hbar}{2\pi mv_F} \int \frac{dx}{x}\sin\left(\frac{mv_F x}{\hbar}\right)\cos\left(\frac{[mu-\hbar k]x}{\hbar}\right), \tag{4.124}$$

where the definition of A and $n_0 = N/L$ were used. In terms of the Fermi length $\lambda_F = \hbar/(mv_F)$,

$$p(k) = \frac{\lambda_F}{2\pi} \int \frac{ds}{s} \sin s \cos\left(\left[k\lambda_F - \frac{u}{v_F}\right]s\right). \tag{4.125}$$

The integral can be evaluated, yielding

$$p(k) = \frac{\lambda_F}{2} \quad \text{for} \quad u - v_F < \frac{\hbar k}{m} < u + v_F,$$

$$p(k) = 0 \quad \text{otherwise}, \tag{4.126}$$

a rather natural result. Moreover, the wavenumber occupation probabilities $p(k)$ satisfy $\int dk\, p(k) = 1$ as they should.

The procedure of this section can be easily extended so as to represent any given equilibrium $f_0(v)$ in terms of a quantum statistical ensemble defined by a set of wavefunctions and the corresponding occupation probabilities. However, to be quantum-mechanically acceptable the conditions (2.56)–(2.59) should be meet so that $f_0(v)$ yield a positive definite density matrix. For the one-dimensional zero-temperature Fermi gas equilibrium, for instance, (2.59) imply $L > \pi\lambda_F$ for the minimal length of the system, otherwise the Heisenberg principle is violated. In other words, the spatial extension of the system should be appreciable larger than the Fermi length, which is a fairly reasonable assumption.

The above mapping from the quantum statistical ensemble to the Wigner function and its inverse can be seen as a partial translation from the quantum-mechanical to the plasma languages and vice versa.

Problems

4.1. Calculate the moments of the Wigner function and demonstrate the expression (4.10) for the pressure in terms of the ensemble wavefunctions.

4.2. By direct calculation demonstrate the expression (4.16) for the pressure as a sum of kinetic, osmotic and quantum contributions.

4.3. Linearize the one-dimensional quantum hydrodynamic model for a zero-temperature Fermi gas to obtain the dispersion relation (4.44).

4.4. Verify that (4.54) and (4.55) cannot sustain small-amplitude, periodic solutions in the $J = 0$ case.

4.5. To calculate the equilibrium Fermi velocity of an electron gas in two and three dimensions, compute the available phase-space hyper-volume, taking into account Pauli's exclusion principle. Deduce (4.101) and (4.102).

4.6. Extend the quantum hydrodynamic model for plasmas, including an energy transport equation.

4.7. Introduce a complex amplitude variable as in Chap. 3 and find a Hamiltonian form for the system (4.54) and (4.55).

4.8. Work out the normalization condition (4.108) in the dilute (nondegenerate) case $\beta \mu \ll 1$, using the properties of the of the polylogarithm function.

4.9. Obtain the projected equilibrium Wigner function in (4.110) for degenerate plasma in the Thomas–Fermi approximation.

4.10. Using the coordinate representation, demonstrate (4.115) for the density matrix in terms of the eigenfunctions of the Hamiltonian operator.

4.11. Demonstrate (4.124).

4.12. Check (4.126).

4.13. Work out condition (2.59) for the one-dimensional zero-temperature Fermi gas equilibrium of Sect. 4.6 to be an acceptable Wigner function.

References

1. Ali, S., Terças, H. and Mendonça J. T.: Nonlocal plasmon excitation in metallic nanostructures. Phys. Rev. B **83**, 153401–153405 (2011)
2. Ancona, M. G., Iafrate, G. J.: Quantum correction to the equation of state of an electron gas in a semiconductor. Phys. Rev. B **39**, 9536–9540 (1989)
3. Ancona, M. G., Tiersten, H. F.: Macroscopic physics of the silicon inversion layer. Phys. Rev. B. **35**, 7959–7965 (1987)
4. Ashcroft, N. W. and Mermin, N. D.: Solid State Physics. Saunders College Publishing, Orlando (1976)
5. Bohm, D. and Hiley, B. J.: The Undivided Universe: an Ontological Interpretation of Quantum Theory. Routledge, London (1993)
6. Bohm, D., Pines, D.: A collective description of electron interactions: III. Coulomb interactions in a degenerate electron gas. Phys. Rev. **92**, 609–625 (1953)
7. Carruthers, P., Zachariasen, F.: Quantum collision theory with phase-space distributions. Rev. Mod. Phys. **55**, 245–285 (1983)
8. Crouseilles, N., Hervieux, P. A. and Manfredi, G.: Quantum hydrodynamic model for the nonlinear electron dynamics in thin metal films. Phys. Rev. B **78**, 155412–155423 (2008)
9. Frensley, W. R.: Boundary conditions for open quantum systems driven far from equilibrium. Rev. Mod. Phys. **62**, 745–791 (1990)
10. Gardner, C.L.: The quantum hydrodynamic model for semiconductor devices. SIAM J. Appl. Math. **54**, 409–427 (1994)
11. Gasser, I., Lin, C. and Markowich, P. A.: A review of dispersive limits of (non)linear Schrödinger-type equations. Taiwan. J. Math. **4**, 501–529 (2000)
12. Ghosh, S. Dubey, S. and Vanshpal, R.: Quantum effect on parametric amplification characteristics in piezoelectric semiconductors. Phys. Lett. A **375**, 43–47 (2010)
13. Haas, F., Manfredi, G., Feix, M. R.: Multistream model for quantum plasmas. Phys. Rev. E **62**, 2763–2772 (2000)

14. Haas, F. Shukla, P. K. and Eliasson, B.: Nonlinear saturation of the Weibel instability in a dense Fermi plasma. J. Plasma Phys. **75**, 251–259 (2009)
15. Haas, F., Manfredi, G., Shukla, P. K. and Hervieux, P.-A.: Breather mode in the many-electron dynamics of semiconductor quantum wells. Phys. Rev. B **80**, 073301–073305 (2009)
16. Kolomeisky, E. B., Newman, T. J., Straley, J. P., Xiaoya, Q.: Low-dimensional Bose liquids: beyond the Gross-Pitaevskii approximation. Phys. Rev. Lett. **85**, 1146–1149 (2000)
17. López, J. L. and Montejo-Gámez, J.: A hydrodynamic approach to multidimensional dissipation-based Schrödinger models from quantum Fokker-Planck dynamics. Physica D **238**, 622–644 (2009)
18. Madelung, E.: Quantum theory in hydrodynamical form. Z. Phys. **40**, 332–336 (1926)
19. Manfredi, G., Feix, M. R.: Theory and simulation of classical and quantum echoes. Phys. Rev. E **53**, 6460–6470 (1996)
20. Manfredi, G., Haas, F.: Self-consistent fluid model for a quantum electron gas. Phys. Rev. B **64**, 075316–075323 (2001)
21. Manfredi, G.: How to model quantum plasmas. Fields Inst. Commun. **46**, 263–287 (2005)
22. Shukla, P. K. and Eliasson, B.: Nonlinear theory for a quantum diode in a dense fermi magnetoplasma. Phys. Rev. Lett. **100**, 036801–036805 (2008)
23. Suh, N. D., Feix, M. R., Bertrand, P.: Numerical simulation of the quantum Liouville–Poisson system. J. Comput Phys. **94**, 403–418 (1991)
24. Wei, L. and Wang, Y. N.: Quantum ion-acoustic waves in single-walled carbon nanotubes studied with a quantum hydrodynamic model. Phys. Rev. B **75**, 193407–193410 (2007)
25. Wigner, E.: On the quantum correction for thermodynamic equilibrium. Phys. Rev. **40**, 749–759 (1932)

Chapter 5
Quantum Ion-Acoustic Waves

Abstract The quantum hydrodynamic model is applied to ion-acoustic waves in quantum plasmas. The corresponding linear dispersion relation is found, as well as a modified Korteweg–de Vries equation for weakly nonlinear solutions. Quantum effects are shown to enlarge the associated solitary wave solutions width. Fully, nonlinear traveling wave solutions are investigated.

5.1 Low Frequency Electrostatic Quantum Plasma Waves

For electrostatic waves with a frequency near the electron plasma frequency, the ion motion is too slow to have any relevance. For this reason, in the previous chapters, the ionic specie was treated as a fixed, homogeneous neutralizing background. However, for slow frequency the ion motion should be taken into account, so that a two species plasma model is needed. By "slow" frequency ω we mean $\omega < \omega_{pi}$, while "high" frequency means $\omega > \omega_{pe}$. Here, $\omega_{pi,pe}$ denote the ion and electron plasma frequencies. In classical electrostatic plasmas, the most important waves are the (high frequency) Langmuir waves and the (low frequency) ion-acoustic waves. In quantum plasmas, we have already discussed a bit about the high frequency modes in Sect. 2.7 resulting on the Bohm–Pines dispersion relation (2.91). Now, it is the time to discuss the quantum counterpart of the classical ion-acoustic modes.

Therefore, following our trend on applications of the quantum plasma hydrodynamic model, let us consider an one-dimensional two species quantum plasma system made by one electronic and one ionic fluid, in the electrostatic approximation [3]. For simplicity, only the continuity and momentum equations will be taken into account, with the energy transport equation ignored. These assumptions are sufficient [4] to describe the classical ion-acoustic wave. Here, we pursue the same methods of classical physics, with only one difference: the

F. Haas, *Quantum Plasmas: An Hydrodynamic Approach*, Springer Series on Atomic, Optical, and Plasma Physics 65, DOI 10.1007/978-1-4419-8201-8_5,
© Springer Science+Business Media, LLC 2011

inclusion of the Bohm term in the momentum equation. Therefore, the relevant equations are given by

$$\frac{\partial n_e}{\partial t} + \frac{\partial (n_e u_e)}{\partial x} = 0, \qquad (5.1)$$

$$\frac{\partial n_i}{\partial t} + \frac{\partial (n_i u_i)}{\partial x} = 0, \qquad (5.2)$$

$$\frac{\partial u_e}{\partial t} + u_e \frac{\partial u_e}{\partial x} = \frac{e}{m_e} \frac{\partial \phi}{\partial x} - \frac{1}{m_e n_e} \frac{\partial P}{\partial x} + \frac{\hbar^2}{2m_e^2} \frac{\partial}{\partial x} \left(\frac{\partial^2 \sqrt{n_e}/\partial x^2}{\sqrt{n_e}} \right), \qquad (5.3)$$

$$\frac{\partial u_i}{\partial t} + u_i \frac{\partial u_i}{\partial x} = -\frac{e}{m_i} \frac{\partial \phi}{\partial x}, \qquad (5.4)$$

$$\frac{\partial^2 \phi}{\partial x^2} = \frac{e}{\varepsilon_0} (n_e - n_i), \qquad (5.5)$$

where $n_{e,i}$ are the electronic and ionic number densities, $u_{e,i}$ the electronic and ionic fluid densities, ϕ the scalar potential, $m_{e,i}$ the electron and ion masses, $-e$ the electron charge, $\hbar = h/(2\pi)$, where h is Planck's constant and ε_0 the vacuum permittivity. Finally, $P = P(n_e)$ is the electron fluid pressure, modeled by some convenient equation of state.

The choice of the equation of state is a delicate subject and depend on the density, temperature, and wavenumber. For definiteness, here the equation of state for a one-dimensional zero-temperature Fermi gas is postulated,

$$P = \frac{m_e v_{Fe}^2}{3n_0^2} n_e^3, \qquad (5.6)$$

where n_0 is the equilibrium density for both electrons and ions, and v_{Fe} is the electrons Fermi velocity, related to the Fermi temperature T_{Fe} by $m_e v_{Fe}^2 = \kappa_B T_{Fe}$, where κ_B is the Boltzmann constant. The results in the following can be trivially modified in the case of other equations of state.

We are disregarding ion pressure terms of both classical and quantum nature. For the electrons, we allow a quantum pressure term due to the Pauli exclusion principle and also a quantum diffraction term which comes from the kinetic term in the Schrödinger equation. In other applications in semiconductor physics, this quantum diffraction term is responsible for tunneling and differential resistance effects [2]. The distinct feature included here is the contribution from Fermi–Dirac statistics.

Equations (5.1) and (5.2) refers to conservation of charge and mass. Equations (5.3) and (5.4) accounts for momentum balance. Equation (5.3), corresponding to balance of the electron fluid momentum, has two quantum terms on the right-hand side. One of them, due to the pressure of the electronic fluid, takes into account the fermionic character of the electrons. The second quantum term, proportional to $\hbar^2$, takes into account the influence of quantum diffraction effects. The ion motion, however, can be taken as classical in view of the high ion mass in comparison to the electron mass. Accordingly, (5.4) contains no quantum terms. Finally, (5.5) is Poisson's equation, describing the self-consistent electrostatic potential.

The electron dynamics can be simplified using general thermodynamic arguments, since electrons reach equilibrium faster than ions due to their smaller mass. Alternatively, let us introduce the following rescaling,

$$\bar{x} = \frac{\omega_{pe} x}{v_{Fe}}, \quad \bar{t} = \omega_{pi} t,$$

$$\bar{n}_e = \frac{n_e}{n_0}, \quad \bar{n}_i = \frac{n_i}{n_0},$$

$$\bar{u}_e = \frac{u_e}{c_s}, \quad \bar{u}_i = \frac{u_i}{c_s}, \quad \bar{\phi} = \frac{e\phi}{\kappa_B T_{Fe}}. \tag{5.7}$$

Here, ω_{pe} and ω_{pi} are the corresponding electron and ion plasma frequencies,

$$\omega_{pe} = \left(\frac{n_0 e^2}{m_e \varepsilon_0}\right)^{1/2}, \quad \omega_{pi} = \left(\frac{n_0 e^2}{m_i \varepsilon_0}\right)^{1/2}. \tag{5.8}$$

Also, c_s is a quantum ion-acoustic velocity, obtained replacing T_e by T_{Fe} in the expression for the classical ion-acoustic velocity,

$$c_s = \left(\frac{\kappa_B T_{Fe}}{m_i}\right)^{1/2}. \tag{5.9}$$

In addition, consider nondimensional parameter

$$H = \frac{\hbar \omega_{pe}}{\kappa_B T_{Fe}}. \tag{5.10}$$

Using the new variables and dropping bars for simplifying notation, we obtain from the electron momentum balance equation (5.3)

$$\frac{m_e}{m_i}\left(\frac{\partial u_e}{\partial t} + u_e \frac{\partial u_e}{\partial x}\right) = \frac{\partial \phi}{\partial x} - n_e \frac{\partial n_e}{\partial x} + \frac{H^2}{2}\frac{\partial}{\partial x}\left(\frac{\partial^2 \sqrt{n_e}/\partial x^2}{\sqrt{n_e}}\right). \tag{5.11}$$

Neglecting the left-hand side of (5.11) due to $m_e/m_i \ll 1$ and considering the boundary conditions $n_e = 1$, $\phi = 0$ at infinity, we obtain

$$\phi = -\frac{1}{2} + \frac{n_e^2}{2} - \frac{H^2}{2\sqrt{n_e}}\frac{d^2}{dx^2}\sqrt{n_e}. \tag{5.12}$$

This last equation is the electrostatic potential in terms of the electron density and its derivatives. If quantum diffraction effects are negligible ($H = 0$), the charge density can be obtained from the potential through an algebraic equation. This is very much like in the classical case where the electron dynamics is simplified assuming the law of altitudes for expressing the electron density in terms of the potential. Here, however, even for $H = 0$ there will be no law of altitudes at all, since the electron equilibrium is given by a Fermi–Dirac distribution and not by a Maxwell–Boltzmann one.

The rescaling (5.7) also implies, from ion conservation of charge and momentum and from Poisson's equation,

$$\frac{\partial n_i}{\partial t} + \frac{\partial (n_i u_i)}{\partial x} = 0, \tag{5.13}$$

$$\frac{\partial u_i}{\partial t} + u_i \frac{\partial u_i}{\partial x} = -\frac{\partial \phi}{\partial x}, \tag{5.14}$$

$$\frac{\partial^2 \phi}{\partial x^2} = n_e - n_i. \tag{5.15}$$

Equations (5.13)–(5.15), together with (5.12), provides a reduced model of four equations with four unknown quantities, n_i, u_i, n_e, and ϕ. This reduced model is the basic tool to be used in the following. The only remaining free parameter is H, which measures the effects of quantum diffraction. Physically, H is the ratio between the electron plasmon energy and the electron Fermi energy.

The reduced model (5.12)–(5.15) support linear waves around the homogeneous equilibrium

$$n_e = n_i = 1, \quad u_i = 0, \quad \phi = 0. \tag{5.16}$$

After linearizing and Fourier-transforming as usual, we find the dispersion relation for these linear waves as

$$\omega^2 = \frac{k^2(1 + H^2 k^2 / 4)}{1 + k^2(1 + H^2 k^2 / 4)}, \tag{5.17}$$

for scaled wave frequency ω and scaled wavenumber k. Assuming small wavenumbers, this gives $\omega = k$, or, reintroducing the original physical variables, a wave propagating at the quantum ion-acoustic velocity c_s as given in the definition (5.9). Equation (5.17) describes the quantum counterpart of the classical ion-acoustic mode, with the Fermi velocity replacing the thermal velocity and with a correction from quantum diffraction effects. Accordingly, we call this new solution the quantum ion-acoustic mode. As for the classical ion-acoustic waves, this mode shows oscillations of both electrons and ions at low frequency.

At the opposite case of small wavelengths, (5.17) gives oscillations at the ion plasma frequency ω_{pi}. This happens because for short wavelengths the electrons are incapable of shielding, so that the ions just oscillate in a background of negative charge. As seen in Fig. 5.1, the asymptotic value $\omega \to \omega_{pi}$ is reached faster for increasing quantum diffraction effects. The reason is that the screening of the electrons become less effective due to the diffusive character of the Bohm potential, responsible for wave packet spreading.

In the following, we investigate the basic properties of the nonlinear solutions for the reduced model (5.12)–(5.15).

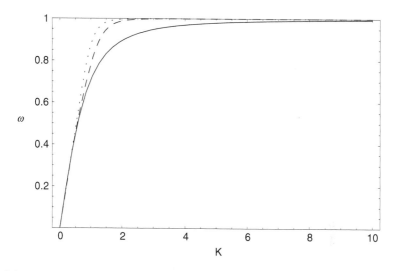

Fig. 5.1 Dispersion relation (5.17) for $H = 0$ (*solid*), $H = 1.5$ (*dashed*) and $H = 3$ (*dotted*)

5.2   A Quantum Korteweg–de Vries Equation

To access weakly nonlinear solutions for the quantum ion-acoustic system (5.12)–(5.15), we use the same singular perturbation methods [1] applied to weakly nonlinear classical waves for an electron–ion plasma. In other words, expand around the equilibrium as

$$n_i = 1 + \varepsilon n_{i1} + \varepsilon^2 n_{i2} + \cdots, \tag{5.18}$$

$$u_i = \varepsilon u_{i1} + \varepsilon^2 u_{i2} + \cdots, \tag{5.19}$$

$$n_e = 1 + \varepsilon n_{e1} + \varepsilon^2 n_{e2} + \cdots, \tag{5.20}$$

where ε is a small parameter proportional to the amplitude of the perturbation. Instead of stopping at first order, the $O(\varepsilon^2)$ terms will be retained.

From (5.12), we have the following expansion for ϕ,

$$\phi = \varepsilon n_{e1} + \frac{\varepsilon^2}{2}\left(n_{e1}^2 + 2n_{e2}\right)$$

$$-\frac{H^2}{4}\frac{\partial^2}{\partial x^2}\left(\varepsilon n_{e1} + \varepsilon^2\left(n_{e2} - \frac{n_{e1}^2}{8}\right)\right) + \frac{H^2 \varepsilon^2 n_{e1}}{8}\frac{\partial^2 n_{e1}}{\partial x^2} + \cdots. \tag{5.21}$$

We use the additional rescaling,

$$\xi = \varepsilon^{1/2}(x - t), \quad \tau = \varepsilon^{3/2} t. \tag{5.22}$$

so that the slow varying time scale is incorporated in the $\sim \varepsilon^{3/2}$ dependence of τ. With these new independent coordinates, one finds from (5.13)–(5.15) and the perturbation expansion (5.18)–(5.21), a set of three equations in the form of power series on ε,

$$\frac{\partial}{\partial \xi}(u_{i1} - n_{i1}) + \varepsilon \left(\frac{\partial n_{i1}}{\partial \tau} + \frac{\partial}{\partial \xi}(-n_{i2} + u_{i2} + n_{i1}u_{i1})\right) = O(\varepsilon^2), \quad (5.23)$$

$$\frac{\partial}{\partial \xi}(n_{e1} - u_{i1}) + \varepsilon \left(\frac{\partial u_{i1}}{\partial \tau} - \frac{\partial u_{i2}}{\partial \xi} + u_{i1}\frac{\partial u_{i1}}{\partial \xi} \right.$$
$$\left. - \frac{H^2}{4}\frac{\partial^3 n_{e1}}{\partial \xi^3} + \frac{1}{2}\frac{\partial}{\partial \xi}(n_{e1}^2 + 2n_{e2})\right) = O(\varepsilon^2) \quad (5.24)$$

$$n_{i1} - n_{e1} + \varepsilon \left(n_{i2} - n_{e2} + \frac{\partial^2 n_{e1}}{\partial \xi^2}\right) = O(\varepsilon^2). \quad (5.25)$$

These three equations should be satisfied to all orders in ε. From the zeroth-order terms plus the boundary condition $u_{i1} \to 0$ at infinity gives

$$n_{e1} = n_{i1} = u_{i1} \equiv U(\xi, \tau), \quad (5.26)$$

in terms of a new function $U(\xi, \tau)$. Equation (5.26) shows that the mode is quasi-neutral in a first approximation. This is a reasonable finding, since for very low frequencies the electrons are viewing an almost stationary ionic background. To prevent strong electric fields, the electron fluid density will try to cancel such quasi-static ionic fluid density. However, due to the inertia, this cancellation will not totally succeed. The deviation from equilibrium then produces the wave's electric field. This reasoning apply to the classical ion-acoustic wave as well [1].

The first-order terms in (5.23)–(5.25), taking into account (5.26), imply

$$\frac{\partial U}{\partial \tau} + \frac{\partial}{\partial \xi}(-n_{i2} + u_{i2} + U^2) = 0, \quad (5.27)$$

$$\frac{\partial U}{\partial \tau} + \frac{\partial}{\partial \xi}\left(-u_{i2} + n_{e2} + U^2 - \frac{H^2}{4}\frac{\partial^2 U}{\partial \xi^2}\right), \quad (5.28)$$

$$\frac{\partial^2 U}{\partial \xi^2} = n_{e2} - n_{i2}. \quad (5.29)$$

Eliminating n_{e2}, n_{i2}, and u_{i2} from (5.27)–(5.29), we obtain a "quantum deformed" Korteweg–de Vries equation,

$$\frac{\partial U}{\partial \tau} + 2U\frac{\partial U}{\partial \xi} + \frac{1}{2}\left(1 - \frac{H^2}{4}\right)\frac{\partial^3 U}{\partial \xi^3} = 0. \quad (5.30)$$

The quantum diffraction effects are responsible for the term proportional to H^2, otherwise we should have the usual Korteweg–de Vries equation.

At first one could imagine that quantum effects would be responsible for the enhancement of the dispersive properties, represented by the third-order derivative term in (5.30). Indeed much of the properties of the Korteweg–de Vries equation, including complete integrability and multi-soliton solutions, follow from the interplay between advection (the $\sim U \partial U / \partial \xi$ term) and dispersion. However, one sees that quantum effects can even invert the sign of the dispersion term, for large enough H. However, this sign is immaterial since we can apply the transformation $\tau \rightarrow -\tau, \xi \rightarrow \xi, U \rightarrow -U$ to change it. Hence, for $H > 2$ the localized solutions (bright solitons) with $U > 0$ of the original equation correspond also to localized solutions, but with inverted polarization ($U < 0$, or dark solitons) and propagating backward in time. A really unexpected feature is that for $H = 2$, the dispersive term disappear. This eventually yields the formation of a shock, much like as for a free ideal neutral classical fluid.

Let us investigate some localized solutions existing for $H \neq 2$. We assume a traveling wave solution of the form

$$U = U(\xi - c\tau), \tag{5.31}$$

for an arbitrary constant phase velocity c. For the sake of definition and without loss of generality, we suppose $c > 0$. Inserting (5.31) into (5.30), integrating once and taking into account the decaying boundary conditions yields

$$\frac{1}{2}\left(1 - \frac{H^2}{4}\right)U'' + U^2 - cU = 0, \tag{5.32}$$

where the prime denotes derivative with respect to $\xi - c\tau$. Equation (5.32) can be rewritten in terms of a Sagdeev effective potential $V(U)$,

$$U'' = -\frac{dV}{dU}, \tag{5.33}$$

where

$$V = \left(1 - \frac{H^2}{4}\right)^{-1}\left(\frac{2U}{3} - c\right)U^2. \tag{5.34}$$

Equation (5.34) admit the energy first integral

$$\Xi = \frac{U'^2}{2} + V(U), \tag{5.35}$$

which, integrated a second time, yields the exact solution in terms of elliptic functions. We restrict ourselves to the analytic solutions in the case of separatrix motion, which is known to be associated with the soliton solution [1]. Figure 5.2 shows

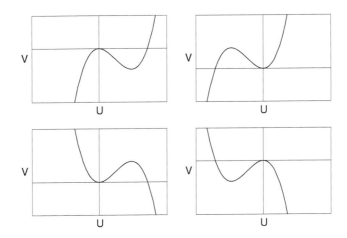

Fig. 5.2 Sagdeev potential in (5.34) for: $H < 2, c > 0$ (*upper, left*); $H < 2, c < 0$ (*upper, right*); $H > 2, c > 0$ (*bottom, left*); $H > 2, c < 0$ (*bottom, right*)

the Sagdeev potential according to H bigger or smaller than 2, and for positive or negative propagation velocity, for the sake of generality. In all cases, $V'(U) = 0$ at $U = 0$ and $U = c$. Also notice that taking $c > 0$ and for $H < 2$, the value of the Sagdeev potential at the local minimum $U = c$ decrease for increasing H. On the other hand, for $H > 2$, the value of the Sagdeev potential at the local maximum also decrease for increasing H. This is a signature that quantum effects tend to destroy localized structures, since for sufficiently large H there will be asymptotically no potential well at all.

For positive c and in the $H < 2$ or semiclassical case, inspection of Fig. 5.2 shows that separatrix motion $\Xi = V(U_{\max})$ at a point of local maximum $U = U_{\max}$ follows from zero energy,

$$\Xi = 0. \tag{5.36}$$

For $\Xi = 0$, integrating (5.35) furnishes

$$U = \frac{3c}{2}\operatorname{sech}^2\left(\sqrt{\frac{c}{2}}\frac{(\xi - c\tau)}{(1 - H^2/4)^{1/2}}\right), \tag{5.37}$$

disregarding an irrelevant integration constant.

For the $H > 2$ or fully quantum case, Fig. 5.2 shows that separatrix motion takes place not for zero energy. Setting $\Xi = V(U_{\max})$ at a point of local maximum $U = U_{\max}$ gives

$$\Xi = \frac{c^3}{3(H^2/4 - 1)}. \tag{5.38}$$

For this value of K, integration of (5.35) furnishes

$$U = c - \frac{3c}{2}\text{sech}^2\left(\sqrt{\frac{c}{2}}\frac{(\xi - c\tau)}{(H^2/4 - 1)^{1/2}}\right), \qquad (5.39)$$

disregarding an irrelevant integration constant.

It is clear from (5.37) and (5.39) that the only influence of quantum effects are on the width of the solitary waves. For both the semiclassical ($H < 2$) and fully quantum ($H > 2$) quantum effects enlarge the solitons widths, which is in accordance with the dispersive property of the Bohm potential. This enlargement is natural in view of wave packet spreading. However, in the nongeneric $H = 2$ case of course no soliton solution can appear, due to the lack of the third-order derivative term in the governing equation.

5.3 Nonlinear Quantum Ion-Acoustic Waves

The developments of the last section are restricted to weakly nonlinearities since we used a second-order expansion on the amplitude parameter. To access arbitrarily large amplitude waves consider traveling wave forms where all quantities are depending only on the variable

$$\zeta = x - Mt, \qquad (5.40)$$

where M is a nondimensional variable playing the rôle of the Mach number of the problem. Equations (5.13) and (5.14) plus the boundary conditions $n_i = 1$, $u_i = \phi = 0$ at infinity gives the conservation laws

$$n_i(u_i - M) = -M, \qquad (5.41)$$

$$\frac{1}{2}(u_i - M)^2 + \phi = \frac{M^2}{2}. \qquad (5.42)$$

Eliminating the ion velocity between (5.41) and (5.42) yields

$$n_i = \left(1 - \frac{2\phi}{M^2}\right)^{-1/2}, \qquad (5.43)$$

expressing the ion density in terms of the electrostatic potential. Defining

$$n_e \equiv A^2 \qquad (5.44)$$

and using (5.12), (6.112), and (5.43), we obtain the dynamical system

$$H^2\frac{d^2A}{d\zeta^2} = A\left(-1+A^4-2\phi\right), \tag{5.45}$$

$$\frac{d^2\phi}{d\zeta^2} = A^2 - \left(1 - \frac{2\phi}{M^2}\right)^{-1/2}. \tag{5.46}$$

Again, we notice the singular character of the formal classical limit $H = 0$, where (5.45) becomes an algebraic equation.

Equations (5.45) and (5.46) are a mechanical system describing the traveling modes of the quantum plasma, depending on the parameters M and H. We are not aware of any exact nonlinear solution for it. Nevertheless, we can find an exact conservation law introducing complex variables in the same way as for (3.25) and (3.26) in the two-stream instability problem. Indeed, with the change of variables

$$\bar{A} = iA, \quad \bar{\phi} = \frac{\phi}{H}, \quad \bar{\zeta} = \frac{\zeta}{H}, \tag{5.47}$$

the system (5.45) and (5.46) takes on the Hamiltonian form

$$\frac{d^2\bar{A}}{d\bar{\zeta}^2} = -\frac{\partial W}{\partial \bar{A}}, \quad \frac{d^2\bar{\phi}}{d\bar{\zeta}^2} = -\frac{\partial W}{\partial \bar{\phi}}, \tag{5.48}$$

in terms of the pseudo-potential $W = W(\bar{A}, \bar{\phi})$ given by

$$W = \frac{\bar{A}^2}{2} - \frac{\bar{A}^6}{6} - M^2\left(1 - \frac{2H\bar{\phi}}{M^2}\right)^{1/2} + H\bar{A}^2\bar{\phi}. \tag{5.49}$$

Hence, there follows immediately the energy first integral

$$I = \frac{1}{2}\left(\frac{d\bar{A}}{d\bar{\zeta}}\right)^2 + \frac{1}{2}\left(\frac{d\bar{\phi}}{d\bar{\zeta}}\right)^2 + W(\bar{A}, \bar{\phi}), \tag{5.50}$$

which, expressed in terms of the original variables, gives the exact constant of motion

$$I = -\frac{H^2}{2}\left(\frac{dA}{d\zeta}\right)^2 + \frac{1}{2}\left(\frac{d\phi}{d\zeta}\right)^2 - \frac{A^2}{2} + \frac{A^6}{6} - M^2\left(1 - \frac{2\phi}{M^2}\right)^{1/2} - A^2\phi. \tag{5.51}$$

Such conservation law can be used to check the accuracy numerical simulations, which should preserve I. However, the constant of motion I cannot be used as a Lyapunov function in nonlinear stability analysis, since their level surfaces are not compact.

The linear stability analysis of (5.45) and (5.46) is worth to be studied. Linearizing around $A = 1$, $\phi = 0$, consider

$$A = 1 + \alpha, \quad \phi = \beta, \tag{5.52}$$

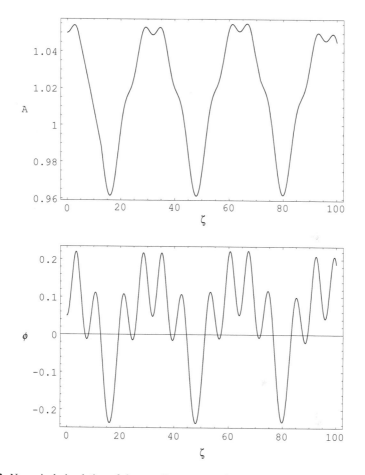

Fig. 5.3 Numerical simulation of the traveling quantum ion-acoustic wave described by (5.45) and (5.46). Parameters: $H = 7$, $M = 1.22$. Initial conditions: $A(0) = 1.05$, $\phi(0) = 0.05$, $A'(0) = 0$, $\phi'(0) = 0$

for small α and β. Retaining only the first-order terms gives

$$\frac{d^2\alpha}{d\zeta^2} = \frac{2}{H^2}(2\alpha - \beta).\tag{5.53}$$

$$\frac{d^2\beta}{d\zeta^2} = 2\alpha - \frac{\beta}{M^2},\tag{5.54}$$

Assuming $\alpha, \beta \sim \exp(ik_c\zeta)$ yields

$$k_c = \frac{1}{2}\left(\frac{4}{H^2} - \frac{1}{J^2} \pm \left(\frac{16}{H^4} + \frac{8}{H^2J^2} + \frac{1}{J^4} - \frac{16}{H^2}\right)^{1/2}\right).\tag{5.55}$$

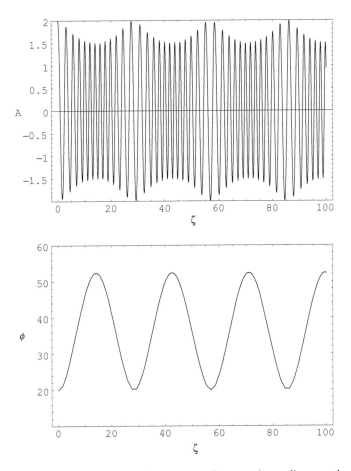

Fig. 5.4 Numerical simulation of the nonlinear quantum ion-acoustic traveling wave described by (5.45) and (5.46). Parameters: $H = 3$, $M = 12$. Initial conditions: $A(0) = 2$, $\phi(0) = 20$, $A'(0) = 0$, $\phi'(0) = 0$

In particular, for

$$H^2 = 4M^2 \tag{5.56}$$

and for supersonic flows ($M^2 > 1$), the system (5.53) and (5.54) shows undamped oscillatory motion with characteristic eigenvalues

$$k_c = \pm \frac{i}{M^2} \sqrt{M^2 - 1}. \tag{5.57}$$

The condition of supersonic flow together with (5.56) shows that this pure oscillatory motion is only possible in the fully quantum regime ($H^2 > 4$).

Numerical simulations can be performed for the nonlinear oscillations of the system (5.45) and (5.46). Typical results are shown in Figs. 5.3 and 5.4.

Problems

5.1. Derive the dispersion relation (5.17), encompassing both quantum ion-acoustic and quantum ion waves.

5.2. Verify the second-order expansion (5.21) for the electrostatic potential.

5.3. Repeat the weakly nonlinear expansion leading to the quantum-mechanically deformed Korteweg–de Vries equation.

5.4. Perform numerical simulations for the nonlinear quantum ion-acoustic waves described by the system (5.45) and (5.46).

References

1. Davidson, R. C.: Methods in Nonlinear Plasma Theory. Academic Press, New York (1972)
2. Gardner, C. L.: The quantum hydrodynamic model for semiconductor devices. SIAM J. Appl. Math. **54**, 409–427 (1994)
3. Haas, F., Garcia, L. G., Goedert, J., Manfredi, G.: Quantum ion acoustic waves. Phys. Plasmas **10**, 3858–3866 (2003)
4. Nicholson, D. R.: Introduction to Plasma Theory. John Wiley, New York (1983)

Chapter 6
Electromagnetic Quantum Plasmas

Abstract The Wigner formalism is extended to systems with magnetic fields. Fluid variables are defined in terms of the electromagnetic Wigner function. The associated evolution equations are shown to include a pressure dyad composed of three parts, corresponding to kinetic velocities dispersion, osmotic velocities dispersion, and a Bohm contribution. With a closure assumption, the quantum counterpart of magnetohydrodynamics is constructed. Exact equilibrium solutions are discussed, showing an oscillatory pattern not present in classical plasma physics.

6.1 Quantum Fluid Equations with Nonzero Magnetic Fields

Until now, we have considered only purely electrostatic quantum plasmas. To include magnetic fields, a first step is to write down the corresponding electromagnetic quantum Vlasov equation. The perspective of this chapter is to consider a gauge-variant formulation, based on the electromagnetic potentials rather than the fields. Later on, in Chap. 9, the model will be improved in terms of a gauge-invariant quantum kinetic equation.

The rather long expression of the electromagnetic Wigner equation provides, in itself, an obvious justification for the use of macroscopic, fluid models. Hence the method of Chap. 4 will be pursued again, considering the Wigner function in terms of the ensemble wavefunctions. Afterwards, the Madelung decomposition of the ensemble wavefunctions allows the identification of the classical and quantum parts of the pressure dyad. At the end, the usual Lorentz force term in the classical electromagnetic fluid equations is recovered. The only difference to the classical equations appears in the addition of a Bohm potential term in the force equation. In comparison to the electrostatic case, the only difference is the addition of the Lorentz force.

Having an electromagnetic quantum fluid model allows for a multitude of developments. For instance, one can: pursue a multi-fluid quantum plasma theory; if charge separation is not so relevant, consider the merging of the negatively and

F. Haas, *Quantum Plasmas: An Hydrodynamic Approach*, Springer Series on Atomic, Optical, and Plasma Physics 65, DOI 10.1007/978-1-4419-8201-8_6,
© Springer Science+Business Media, LLC 2011

positively charged species. More exactly, consider only the global properties of the system, building the quantum analog of the classical magnetohydrodynamic equations. The resulting quantum magnetohydrodynamic model was presented in [10], being extended for inclusion of spin variables in [5].

To begin with, one needs the kinetic equation satisfied by the Wigner function. To simplify notation, it is easier to consider a single species plasma, with the extension to multi-species plasmas being straightforward. Further, in the context of a mean field theory, it suffices to deal with one-particle wavefunctions. In other words, the full N-body wavefunction is assumed to be in the factorized form of a product of equal one-body wavefunctions. So consider a statistical mixture with M states $\psi_\alpha = \psi_\alpha(\mathbf{r},t)$, each with probability p_α, with $\alpha = 1,\ldots,M$, so that $p_\alpha \geq 0$ and $\sum_{\alpha=1}^{M} p_\alpha = 1$. The one-body wavefunctions satisfy Schrödinger's equation,

$$\frac{1}{2m}(-i\hbar\nabla - q\mathbf{A})^2\,\psi_\alpha + q\phi\,\psi_\alpha = i\hbar\frac{\partial\psi_\alpha}{\partial t}. \tag{6.1}$$

Here, we consider charge carriers of mass m and charge q, subjected to mean field self-consistent scalar and vector potentials $\phi = \phi(\mathbf{r},t)$ and $\mathbf{A} = \mathbf{A}(\mathbf{r},t)$, respectively. For definiteness, the Coulomb gauge $\nabla \cdot \mathbf{A} = 0$ is chosen.

In the electromagnetic case, it is more customary to define the Wigner function in phase space in terms of position $\mathbf{r}$ and momentum $\mathbf{p} = m\mathbf{v} + q\mathbf{A}$. Then, the Wigner function $f = f(\mathbf{r},\mathbf{p},t)$ is defined [6] as

$$f(\mathbf{r},\mathbf{p},t) = \frac{N}{(2\pi\hbar)^3}\sum_{\alpha=1}^{M} p_\alpha \int d\mathbf{s}\,\psi_\alpha^*\left(\mathbf{r}+\frac{\mathbf{s}}{2},t\right) e^{\frac{i\mathbf{p}\cdot\mathbf{s}}{\hbar}}\,\psi_\alpha\left(\mathbf{r}-\frac{\mathbf{s}}{2},t\right), \tag{6.2}$$

where N is the number of charge carriers and the ensemble wavefunctions ψ_α have unit norm. Since the wavefunctions are changing in time, the same apply to $f = f(\mathbf{r},\mathbf{p},t)$. A long calculation yields the following integro-differential equation,

$$\frac{\partial f}{\partial t} + \frac{\mathbf{p}}{m}\cdot\nabla f$$

$$= \frac{iq}{\hbar(2\pi\hbar)^3}\iint d\mathbf{s}\,d\mathbf{p}'\,e^{\frac{i(\mathbf{p}-\mathbf{p}')\cdot\mathbf{s}}{\hbar}}\left[\phi\left(\mathbf{r}+\frac{\mathbf{s}}{2},t\right) - \phi\left(\mathbf{r}-\frac{\mathbf{s}}{2},t\right)\right]f(\mathbf{r},\mathbf{p}',t)$$

$$+ \frac{iq^2}{2\hbar m(2\pi\hbar)^3}\iint d\mathbf{s}\,d\mathbf{p}'\,e^{\frac{i(\mathbf{p}-\mathbf{p}')\cdot\mathbf{s}}{\hbar}}\left[A^2\left(\mathbf{r}+\frac{\mathbf{s}}{2},t\right) - A^2\left(\mathbf{r}-\frac{\mathbf{s}}{2},t\right)\right]f(\mathbf{r},\mathbf{p}',t)$$

$$+ \frac{q}{2m(2\pi\hbar)^3}\nabla\cdot\iint d\mathbf{s}\,d\mathbf{p}'\,e^{\frac{i(\mathbf{p}-\mathbf{p}')\cdot\mathbf{s}}{\hbar}}\left[\mathbf{A}\left(\mathbf{r}+\frac{\mathbf{s}}{2},t\right) + \mathbf{A}\left(\mathbf{r}-\frac{\mathbf{s}}{2},t\right)\right]f(\mathbf{r},\mathbf{p}',t)$$

$$- \frac{iq}{\hbar m(2\pi\hbar)^3}\mathbf{p}\cdot\iint d\mathbf{s}\,d\mathbf{p}'\,e^{\frac{i(\mathbf{p}-\mathbf{p}')\cdot\mathbf{s}}{\hbar}}\left[\mathbf{A}\left(\mathbf{r}+\frac{\mathbf{s}}{2},t\right) - \mathbf{A}\left(\mathbf{r}-\frac{\mathbf{s}}{2},t\right)\right]f(\mathbf{r},\mathbf{p}',t). \tag{6.3}$$

Let us sketch the derivation of (6.3), which has to be compared with (4.1) valid in the electrostatic case. In the Coulomb gauge, one has

$$-\frac{\hbar^2}{2m}\nabla^2\psi_\alpha + \left(q\phi + \frac{q^2A^2}{2m}\right)\psi_\alpha + \frac{i\hbar q}{m}\mathbf{A}\cdot\nabla\psi_\alpha = i\hbar\frac{\partial\psi_\alpha}{\partial t}. \tag{6.4}$$

By inspection apart from the $\sim\mathbf{A}\cdot\nabla\psi_\alpha$ the electromagnetic case could be recovered from the electrostatic case taking

$$q\phi \to q\phi + \frac{q^2A^2}{2m}, \tag{6.5}$$

in (4.1), together with the *formal* replacement $\mathbf{v} \to \mathbf{p}/m$, even though actually momentum $\mathbf{p}$ and velocity $\mathbf{v}$ are related by $\mathbf{p} = m\mathbf{v} + q\mathbf{A}$. This gives the first and second terms in the right-hand side of (6.3). The remaining contributions are the third and fourth terms in the right-hand side of (6.3). According to the reasoning, the origin of them is just on the

$$\frac{\partial\psi_\alpha}{\partial t} = \frac{q}{m}\mathbf{A}\cdot\nabla\psi_\alpha + \cdots \tag{6.6}$$

part of Schrödinger's equation. To proceed, one needs the identity

$$\int d\mathbf{s}\, e^{\frac{i\mathbf{p}\cdot\mathbf{s}}{\hbar}} \psi_\alpha^*\left(\mathbf{r}+\frac{\mathbf{s}}{2},t\right) \mathbf{A}\left(\mathbf{r}+\frac{\mathbf{s}}{2},t\right)\cdot\nabla\psi_\alpha\left(\mathbf{r}-\frac{\mathbf{s}}{2},t\right)$$
$$+ \int d\mathbf{s}\, e^{\frac{i\mathbf{p}\cdot\mathbf{s}}{\hbar}} \mathbf{A}\left(\mathbf{r}-\frac{\mathbf{s}}{2},t\right)\cdot\nabla\psi_\alpha^*\left(\mathbf{r}+\frac{\mathbf{s}}{2},t\right) \psi_\alpha\left(\mathbf{r}-\frac{\mathbf{s}}{2},t\right)$$
$$= \frac{i\mathbf{p}}{\hbar}\cdot\int d\mathbf{s}\, e^{\frac{i\mathbf{p}\cdot\mathbf{s}}{\hbar}} \left(\mathbf{A}\left(\mathbf{r}+\frac{\mathbf{s}}{2},t\right) - \mathbf{A}\left(\mathbf{r}-\frac{\mathbf{s}}{2},t\right)\right) \psi_\alpha^*\left(\mathbf{r}+\frac{\mathbf{s}}{2},t\right) \psi_\alpha\left(\mathbf{r}-\frac{\mathbf{s}}{2},t\right), \tag{6.7}$$

valid in the Coulomb gauge and proved after a couple of integrations by parts considering

$$\frac{\partial\psi_\alpha}{\partial\mathbf{s}}\left(\mathbf{r}-\frac{\mathbf{s}}{2},t\right) = -\frac{1}{2}\nabla\psi_\alpha\left(\mathbf{r}-\frac{\mathbf{s}}{2},t\right), \tag{6.8}$$

$$\frac{\partial\psi_\alpha^*}{\partial\mathbf{s}}\left(\mathbf{r}+\frac{\mathbf{s}}{2},t\right) = \frac{1}{2}\nabla\psi_\alpha^*\left(\mathbf{r}+\frac{\mathbf{s}}{2},t\right). \tag{6.9}$$

In addition, the following identity is necessary,

$$\int d\mathbf{p} f(\mathbf{r},\mathbf{p},t) e^{-\frac{i\mathbf{p}\cdot\mathbf{s}}{\hbar}} = N\sum_{\alpha=1}^{M} p_\alpha \psi_\alpha^*\left(\mathbf{r}+\frac{\mathbf{s}}{2},t\right) \psi_\alpha\left(\mathbf{r}\frac{\mathbf{s}}{2},t\right). \tag{6.10}$$

Taking into account (6.6)–(6.10) one arrives at the electromagnetic quantum Vlasov equation (6.3).

Notice that the Wigner function in Eq. (6.2) is *not* invariant under the gauge transformations

$$\phi \to \phi - \frac{\partial \Lambda}{\partial t}, \quad \mathbf{A} \to \mathbf{A} + \nabla \Lambda, \quad \psi_\alpha \to \psi_\alpha \exp\left(\frac{iq\Lambda}{\hbar}\right), \quad (6.11)$$

where $\Lambda = \Lambda(\mathbf{r}, t)$ is an arbitrary differentiable function. We postpone a more serious discussion on gauge invariance issues to Chap. 11. For the derivation of electromagnetic quantum plasma equations and quantum magnetohydrodynamics, the present gauge-dependent formulation is enough and gives faithful prescriptions.

In the formal classical limit ($\hbar \to 0$), (6.3) becomes the Vlasov equation,

$$\frac{\partial f}{\partial t} + \mathbf{v} \cdot \nabla f + \frac{q}{m}(\mathbf{E} + \mathbf{v} \times \mathbf{B}) \cdot \frac{\partial f}{\partial \mathbf{v}} = 0, \quad (6.12)$$

where $\mathbf{v} = (\mathbf{p} - q\mathbf{A})/m$, $\mathbf{E} = -\nabla \phi - \partial \mathbf{A}/\partial t$ and $\mathbf{B} = \nabla \times \mathbf{A}$. The proof is as follows: substitute $\mathbf{s} = \hbar \mathbf{S}$ in (6.3) so as to safely expanding in powers of $\hbar$ without any singularity; take the change of variables

$$T = t, \quad \mathbf{R} = \mathbf{r}, \quad \mathbf{v} = \frac{1}{m}(\mathbf{p} - q\mathbf{A}) \quad (6.13)$$

together with

$$\frac{\partial}{\partial t} = \frac{\partial}{\partial T} - \frac{q}{m}\frac{\partial \mathbf{A}}{\partial t} \cdot \frac{\partial}{\partial \mathbf{v}}, \quad (6.14)$$

$$\frac{\partial}{\partial r_i} = \frac{\partial}{\partial R_i} - \frac{q}{m}\sum_{j=1}^{3}\frac{\partial A_j}{\partial r_i}\frac{\partial}{\partial v_j}, \quad i = 1, 2, 3, \quad (6.15)$$

$$\frac{\partial}{\partial \mathbf{p}} = \frac{1}{m}\frac{\partial}{\partial \mathbf{v}} \quad (6.16)$$

Retaining just the classical contribution in the Taylor expansion of (6.3) in powers of $\hbar$ gives Vlasov's equation (6.12), at the end replacing back $T = t, \mathbf{R} = \mathbf{r}$.

The electromagnetic Wigner equation is written in terms of the mean field potentials, with the inclusion of external fields being easily done if necessary. The determination of $\mathbf{A}, \phi$ requires Maxwell equations, where the charge and current densities are given by moments of the Wigner function. In this way, one arrives at a self-consistent kinetic description for electromagnetic quantum plasmas. Equation (6.3) has been first obtained, with a different notation, in [1] and rediscovered, in the case of homogeneous magnetic fields, in [14]. The concept of electromagnetic Wigner function has been addressed long ago in [8], but without the explicit derivation of the corresponding evolution equation. Applications of spin-dependent Wigner functions in spintronics were done in [16]. Relativistic Wigner functions depending on spin degrees of freedom were addressed in [13].

With probably some exceptions, (6.3) is too complicated to be of any use. Hence, as in Chap. 4, a fluid approach is advisable. The resulting quantum hydromagnetic model share the same limitations of the electrostatic quantum hydrodynamic model for plasmas. In particular, no kinetic effects can be described within such framework. Nevertheless, to have sufficiently tractable equations, let us introduce the following moments of the Wigner function: the fluid density

$$n = \int d\mathbf{p}\, f, \tag{6.17}$$

the fluid velocity

$$\mathbf{u} = \frac{1}{mn} \int d\mathbf{p}\, (\mathbf{p} - q\mathbf{A})\, f \tag{6.18}$$

and the pressure dyad

$$\mathbf{P} = \frac{1}{m} \int d\mathbf{p}\, (\mathbf{p} - q\mathbf{A}) \otimes (\mathbf{p} - q\mathbf{A})\, f - mn\mathbf{u} \otimes \mathbf{u}. \tag{6.19}$$

Proceeding as in Chap. 4, but now using (6.3), the result is

$$\frac{\partial n}{\partial t} + \nabla \cdot (n\mathbf{u}) = 0, \tag{6.20}$$

$$\frac{\partial \mathbf{u}}{\partial t} + \mathbf{u} \cdot \nabla \mathbf{u} = -\frac{1}{mn} \nabla \cdot \mathbf{P} + \frac{q}{m}(\mathbf{E} + \mathbf{u} \times \mathbf{B}). \tag{6.21}$$

Equations (6.20) and (6.21) are the formally same as the classical fluid equations since there's no trace of Planck's constant in it. However, first express the fluid variables in terms of the wavefunctions. With the aid of the identities

$$\mathbf{p}\exp\left(\frac{i\mathbf{p} \cdot \mathbf{s}}{\hbar}\right) = -i\hbar \frac{\partial}{\partial \mathbf{s}} \exp\left(\frac{i\mathbf{p} \cdot \mathbf{s}}{\hbar}\right), \tag{6.22}$$

$$\mathbf{p} \otimes \mathbf{p}\exp\left(\frac{i\mathbf{p} \cdot \mathbf{s}}{\hbar}\right) = -\hbar^2 \left(\frac{\partial}{\partial \mathbf{s}}\right) \otimes \left(\frac{\partial}{\partial \mathbf{s}}\right) \exp\left(\frac{i\mathbf{p} \cdot \mathbf{s}}{\hbar}\right), \tag{6.23}$$

we find

$$n = N \sum_{\alpha=1}^{M} p_\alpha |\psi_\alpha|^2, \tag{6.24}$$

$$\mathbf{u} = \frac{i\hbar N}{2mn} \sum_{\alpha=1}^{M} p_\alpha (\psi_\alpha \nabla \psi_\alpha^* - \psi_\alpha^* \nabla \psi_\alpha) - \frac{q}{m}\mathbf{A}, \tag{6.25}$$

$$\mathbf{P} = -\frac{\hbar^2 N}{4m} \sum_{\alpha=1}^{M} p_\alpha \left[(\nabla \otimes \nabla)(|\psi_\alpha|^2) - 2(\nabla \psi_\alpha^*) \otimes (\nabla \psi_\alpha) - 2(\nabla \psi_\alpha) \otimes (\nabla \psi_\alpha^*)\right]$$

$$+ \frac{\hbar^2 N^2}{4mn} \sum_{\alpha,\beta=1}^{M} p_\alpha p_\beta (\psi_\alpha \nabla \psi_\alpha^* - \psi_\alpha^* \nabla \psi_\alpha) \otimes (\psi_\beta \nabla \psi_\beta^* - \psi_\beta^* \nabla \psi_\beta). \tag{6.26}$$

As in Chap. 4, consider the Madelung decomposition

$$\psi_\alpha = \sqrt{n_\alpha}\, e^{iS_\alpha/\hbar}, \quad \alpha = 1,\ldots,M, \tag{6.27}$$

for real functions $n_\alpha = n_\alpha(\mathbf{r},t)$ and $S_\alpha = S_\alpha(\mathbf{r},t)$. Equations (6.24)–(6.26) then gives

$$n = N \sum_{\alpha=1}^M p_\alpha n_\alpha, \tag{6.28}$$

$$\mathbf{u} = \frac{N}{mn} \sum_{\alpha=1}^M p_\alpha n_\alpha \nabla S_\alpha - \frac{q}{m}\mathbf{A}, \tag{6.29}$$

$$\mathbf{P} = \frac{N}{m} \sum_{\alpha=1}^M p_\alpha n_\alpha \nabla S_\alpha \otimes \nabla S_\alpha - \frac{N^2}{mn} \sum_{\alpha,\beta=1}^M p_\alpha p_\beta n_\alpha n_\beta \nabla S_\alpha \otimes \nabla S_\beta$$

$$+ \frac{\hbar^2 N}{4m} \sum_{\alpha=1}^M p_\alpha \frac{\nabla n_\alpha \otimes \nabla n_\alpha}{n_\alpha} - \frac{\hbar^2 N}{4m} \sum_{\alpha=1}^M p_\alpha \nabla \otimes \nabla n_\alpha. \tag{6.30}$$

We can see the explicit presence of Planck's constant only at the final two terms of the second moment of the Wigner function. At first glance, we could identify these $\hbar$-dependent terms as the "quantum" part of the pressure dyad. This terminology would reflect the fact that in a semiclassical expansion the leading contribution would be entirely in the $\hbar$-independent terms in $\mathbf{P}$. However, the pressure dyad can be rewritten [9] as a sum of: an average dispersion of the usual velocities $\mathbf{u}_\alpha$ defined as

$$\mathbf{u}_\alpha = \frac{1}{m}[\nabla S_\alpha - q\mathbf{A}], \quad \alpha = 1,\ldots,M, \tag{6.31}$$

an average dispersion of the osmotic velocities $\mathbf{u}_\alpha^o$ defined as

$$\mathbf{u}_\alpha^o = \frac{\hbar}{2m} \frac{\nabla n_\alpha}{n_\alpha}, \quad \alpha = 1,\ldots,M, \tag{6.32}$$

and a Bohm potential term. Indeed, direct calculation shows that (6.30) is equivalent to

$$\mathbf{P} = \frac{mn}{2} \sum_{\alpha,\beta=1}^M \tilde{p}_\alpha \tilde{p}_\beta \left[(\mathbf{u}_\alpha - \mathbf{u}_\beta) \otimes (\mathbf{u}_\alpha - \mathbf{u}_\beta) + (\mathbf{u}_\alpha^o - \mathbf{u}_\beta^o) \otimes (\mathbf{u}_\alpha^o - \mathbf{u}_\beta^o) \right]$$

$$- \frac{\hbar^2 n}{4m} \nabla \otimes \nabla \ln n, \tag{6.33}$$

with modified statistical weights $\tilde{p}_\alpha$ given by

$$\tilde{p}_\alpha = \frac{N p_\alpha n_\alpha}{n}, \quad \alpha = 1,\ldots,M, \tag{6.34}$$

satisfying $\tilde{p}_\alpha \geq 0, \sum_{\alpha=1}^M \tilde{p}_\alpha = 1$. In the three-dimensional space, we can interpret the osmotic velocities more properly than in Chap. 4, since it is apparent that they are directed to the regions of highest densities. The terminology "osmotic" stems from the fact that $\mathbf{u}_\alpha^o = 0$ for zero density gradients [4].

Since the two first terms in the right-hand side of (6.33) are a sum of ordinary and osmotic velocity dispersions, we can identify

$$\mathbf{P}^C = \frac{mn}{2} \sum_{\alpha,\beta=1}^M \tilde{p}_\alpha \tilde{p}_\beta \left[(\mathbf{u}_\alpha - \mathbf{u}_\beta) \otimes (\mathbf{u}_\alpha - \mathbf{u}_\beta) + (\mathbf{u}_\alpha^o - \mathbf{u}_\beta^o) \otimes (\mathbf{u}_\alpha^o - \mathbf{u}_\beta^o) \right] \quad (6.35)$$

as the classical part of the pressure dyad, while the Bohm potential term

$$\mathbf{P}^Q = -\frac{\hbar^2 n}{4m} \nabla \otimes \nabla \ln n \quad (6.36)$$

can be identified as the intrinsically quantum part of the pressure dyad, so that $\mathbf{P} = \mathbf{P}^C + \mathbf{P}^Q$.

Some comments are in order here: (a) in the classical limit, only $\mathbf{P}^C$ survives, moreover with a vanishing osmotic pressure contribution; (b) if we consider an extended random phase approximation, in the sense of taking a statistical ensemble with wavefunctions all with the same amplitude, or

$$\psi_\alpha = \sqrt{\frac{n}{N}} \, e^{iS_\alpha/\hbar}, \quad \alpha = 1, \dots, M, \quad (6.37)$$

then the classical pressure would be entirely contained in the ordinary velocities dispersion; (c) mathematically from (6.36) the Bohm potential is equivalent to a nondiagonal pressure dyad. However, it is not a pressure in the thermodynamic sense, because it has nothing to do with a measure of the velocities fluctuation. Indeed, $\mathbf{P}^Q$ is nonzero even in the pure state case. Rather, the Bohm potential should be understood as a consequence from the wave-like, dispersive properties of the quantum fluid.

To proceed, some closure assumption should be made. Since $\mathbf{P}^C$ is written as the sum of average velocity dispersions, it is natural to define a diagonal, isotropic form where the components P_{ij}^C of the classical pressure dyad are given by

$$P_{ij}^C = \delta_{ij} P^C, \quad (6.38)$$

where the scalar pressure $P^C = P^C(n)$ represents a suitable equation of state. Strong magnetic fields have to be treated more carefully, since they are associated with anisotropic pressure dyads. Such possibility will not be pursued here, for simplicity.

With the assumed closure, the momentum transport equation (6.21) becomes

$$\frac{\partial \mathbf{u}}{\partial t} + \mathbf{u} \cdot \nabla \mathbf{u} = -\frac{1}{mn} \nabla P^C(n) + \frac{q}{m}(\mathbf{E} + \mathbf{u} \times \mathbf{B}) + \frac{\hbar^2}{2m^2} \nabla \left(\frac{\nabla^2 \sqrt{n}}{\sqrt{n}} \right), \quad (6.39)$$

which is just the classical fluid equation modified by the extra $\sim\hbar^2$ term, or Bohm potential term. The equation of continuity (6.20) and the force equation (6.39), together with Maxwell equations, constitute a quantum hydrodynamic model for magnetized plasmas.

6.2   Quantum Magnetohydrodynamics

It is rather direct to generalize the previous model to two-species plasmas. In general, for a multi-species plasma, one starts with a Wigner function for each species. Calculating moments as before, the result is a set of hydrodynamic equations for each particular specie. Consider an electron fluid with number density n_e, fluid velocity $\mathbf{u}_e$, charge $-e$, mass m_e and scalar pressure P_e. Similarly, take ions with number density n_i, fluid velocity $\mathbf{u}_i$, charge e, mass m_i and pressure P_i. Presently, we get the following bipolar quantum fluid model,

$$\frac{\partial n_e}{\partial t} + \nabla \cdot (n_e \mathbf{u}_e) = 0, \tag{6.40}$$

$$\frac{\partial n_i}{\partial t} + \nabla \cdot (n_i \mathbf{u}_i) = 0, \tag{6.41}$$

$$\frac{\partial \mathbf{u}_e}{\partial t} + \mathbf{u}_e \cdot \nabla \mathbf{u}_e = -\frac{\nabla P_e}{m_e n_e} - \frac{e}{m_e}(\mathbf{E} + \mathbf{u}_e \times \mathbf{B})$$
$$+ \frac{\hbar^2}{2m_e^2} \nabla \left(\frac{\nabla^2 \sqrt{n_e}}{\sqrt{n_e}} \right) - \nu_{ei}(\mathbf{u}_e - \mathbf{u}_i), \tag{6.42}$$

$$\frac{\partial \mathbf{u}_i}{\partial t} + \mathbf{u}_i \cdot \nabla \mathbf{u}_i = -\frac{\nabla P_i}{m_i n_i} + \frac{e}{m_i}(\mathbf{E} + \mathbf{u}_i \times \mathbf{B})$$
$$+ \frac{\hbar^2}{2m_i^2} \nabla \left(\frac{\nabla^2 \sqrt{n_i}}{\sqrt{n_i}} \right) - \nu_{ie}(\mathbf{u}_i - \mathbf{u}_e). \tag{6.43}$$

In (6.42) and (6.43), we have added some often used phenomenological collision terms, with the collision frequencies ν_{ei} and ν_{ie} accounting for the momentum transfer between the electronic and ionic fluids. Global momentum conservation in collisions imply $m_e \nu_{ei} = m_i \nu_{ie}$, so that $\nu_{ie} \ll \nu_{ei}$ since typically the ions are much more massive than electrons. In addition, notice that a first principles derivation would consider dissipative quantum mechanics, a rather controversial field. Finally, the exact form of the collision terms is not essential, as long as global momentum conservation is assured [3, 15].

Equations (6.40)–(6.43) have to be supplemented by Maxwell equations,

$$\nabla \cdot \mathbf{E} = \frac{\rho}{\varepsilon_0}, \tag{6.44}$$

$$\nabla \cdot \mathbf{B} = 0, \tag{6.45}$$

$$\nabla \times \mathbf{E} = -\frac{\partial \mathbf{B}}{\partial t}, \tag{6.46}$$

$$\nabla \times \mathbf{B} = \mu_0 \mathbf{J} + \mu_0 \varepsilon_0 \frac{\partial \mathbf{E}}{\partial t}, \tag{6.47}$$

where the charge and current densities are given, respectively, by

$$\rho = e(n_i - n_e), \quad \mathbf{J} = e(n_i \mathbf{u}_i - n_e \mathbf{u}_e). \tag{6.48}$$

Equations (6.40)–(6.48) are an electromagnetic quantum hydrodynamic model for a two-species plasma.

One can directly work using (6.40)–(6.48), generalizing the classical two-fluid plasma approach. However, often one is not so interested in the behavior of each species. Rather a more rough, but nevertheless very useful approach is to study the properties of the entire plasma, in terms of an one-fluid model. We chose this last view, proceeding in entire analogy with the classical magnetohydrodynamic modeling. Hence, define the global mass density

$$\rho_m = m_e n_e + m_i n_i \tag{6.49}$$

and the global fluid velocity

$$\mathbf{U} = \frac{m_e n_e \mathbf{u}_e + m_i n_i \mathbf{u}_i}{m_e n_e + m_i n_i}. \tag{6.50}$$

The electronic and ionic densities are defined in terms of the global mass and charge densities according to

$$n_e = \frac{1}{m_i + m_e} \left(\rho_m - \frac{m_i}{e} \rho \right), \tag{6.51}$$

$$n_i = \frac{1}{m_i + m_e} \left(\rho_m + \frac{m_e}{e} \rho \right). \tag{6.52}$$

It is also useful to express each species fluid velocity in terms of the one-fluid velocity and the current density, by means of

$$\left(\rho_m - \frac{m_i}{e} \rho \right) \mathbf{u}_e = \rho_m \mathbf{U} - \frac{m_i}{e} \mathbf{J}, \tag{6.53}$$

$$\left(\rho_m + \frac{m_e}{e} \rho \right) \mathbf{u}_i = \rho_m \mathbf{U} + \frac{m_e}{e} \mathbf{J}. \tag{6.54}$$

We desire the transport equations for the global, one-fluid quantities. Multiplying the continuity equations (6.40) and (6.41), respectively, by m_e and m_i and adding the results, we obtain a continuity equation for the global mass density. In the same manner, multiplying the force equations (6.42) and (6.43) by $m_e n_e$ and $m_i n_i$, respectively, and adding the results, we obtain a transport equation for the global velocity field. Specifically,

$$\frac{\partial \rho_m}{\partial t} + \nabla \cdot (\rho_m \mathbf{U}) = 0, \tag{6.55}$$

$$\rho_m \left(\frac{\partial \mathbf{U}}{\partial t} + \mathbf{U} \cdot \nabla \mathbf{U} \right) = -\nabla \cdot \tilde{\mathbf{P}} + \mathbf{J} \times \mathbf{B} + \frac{\hbar^2 \rho_m}{2 m_e m_i} \nabla \left(\frac{\nabla^2 \sqrt{\rho_m}}{\sqrt{\rho_m}} \right), \tag{6.56}$$

where quasi-neutrality ($n_e = n_i, \rho = 0$) was used. In addition, an one-fluid pressure was defined,

$$\tilde{\mathbf{P}} = P\mathbf{I} + \frac{m_e m_i n_e n_i}{\rho_m} (\mathbf{u}_e - \mathbf{u}_i) \otimes (\mathbf{u}_e - \mathbf{u}_i), \tag{6.57}$$

with $P = P_e + P_i$ and where $\mathbf{I}$ is the identity matrix. Further, supposing $P_e = P_i = P/2$ and disregarding the last term in (6.57), which is reasonable except for very large current densities, we obtain

$$\frac{\partial \mathbf{U}}{\partial t} + \mathbf{U} \cdot \nabla \mathbf{U} = -\frac{1}{\rho_m} \nabla P + \frac{1}{\rho_m} \mathbf{J} \times \mathbf{B} + \frac{\hbar^2}{2 m_e m_i} \nabla \left(\frac{\nabla^2 \sqrt{\rho_m}}{\sqrt{\rho_m}} \right). \tag{6.58}$$

In the quasi-neutral case, we need the transport equations only for ρ_m, $\mathbf{U}$ and $\mathbf{J}$. Hence, consider now the current density. While (6.55) and (6.56) are valid regardless of the mass of the ions, from now on it is preferable to assume $m_e \ll m_i$ to be allowed to neglect a large number of terms. This is a reasonable assumption, for example, for hydrogen plasma ($m_i \simeq 1,836 m_e$). Then, in the combined quasi-neutral and $m_e/m_i \ll 1$ limits, we get the following equation for the current density $\mathbf{J}$,

$$\frac{m_e m_i}{\rho_m e^2} \frac{\partial \mathbf{J}}{\partial t} - \frac{m_i \nabla P_e}{\rho_m e} = \mathbf{E} + \mathbf{U} \times \mathbf{B} - \frac{m_i}{\rho_m e} \mathbf{J} \times \mathbf{B} - \frac{\hbar^2}{2 e m_e} \nabla \left(\frac{\nabla^2 \sqrt{\rho_m}}{\sqrt{\rho_m}} \right) - \frac{1}{\sigma} \mathbf{J}, \tag{6.59}$$

where $\sigma = \rho_m e^2 / (m_e m_i \nu_{ei})$ is the longitudinal electrical conductivity. In the derivation, all nonlinear terms involving either $\mathbf{U}$ and $\mathbf{J}$ were disregarded, as usual in magnetohydrodynamics [3]. This is the reason why advective terms like $\mathbf{U} \cdot \nabla \mathbf{J}$ are not present in (6.59). For strongly turbulent plasma, the approximation certainly breaks down. In this same equation because the current density is regarded as a small perturbation one can consider σ as a constant, evaluated at the equilibrium value of the mass density. Also $\nu_{ie} \ll \nu_{ei}$ was taken into account.

Equation (6.59) is the quantum version of the generalized Ohm's law [3, 15] because neglecting magnetic fields, quantum effects, and the left-hand side of it, we obtain $\mathbf{J} = \sigma\mathbf{E}$. The continuity equation (6.55), the force equation (6.58), the quantum version of the generalized Ohm's law (6.59), an equation of state for the scalar pressure, plus Maxwell equations, provides a closed system of equations. However, it can be further simplified, assuming a very large conductivity and a slowly varying time-scale.

6.3 Simplified and Ideal Quantum Magnetohydrodynamic Models

Usually [3, 15], the left-hand side of the equation (6.59) is neglected in the cases of slowly varying processes and small pressures. Also, for slowly varying and high conductivity problems, the displacement current can be neglected in Ampère's law. Finally, for definiteness, we assume an equation of state appropriated for adiabatic processes. In this way, we get a set of simplified quantum magnetohydrodynamic equations:

$$\frac{\partial \rho_m}{\partial t} + \nabla \cdot (\rho_m \mathbf{U}) = 0, \tag{6.60}$$

$$\frac{\partial \mathbf{U}}{\partial t} + \mathbf{U} \cdot \nabla \mathbf{U} = -\frac{1}{\rho_m}\nabla P + \frac{1}{\rho_m}\mathbf{J} \times \mathbf{B} + \frac{\hbar^2}{2m_e m_i}\nabla\left(\frac{\nabla^2\sqrt{\rho_m}}{\sqrt{\rho_m}}\right), \tag{6.61}$$

$$\nabla P = V_s^2 \nabla \rho_m, \tag{6.62}$$

$$\nabla \times \mathbf{E} = -\frac{\partial \mathbf{B}}{\partial t}, \tag{6.63}$$

$$\nabla \times \mathbf{B} = \mu_0 \mathbf{J}, \tag{6.64}$$

$$\mathbf{J} = \sigma\left[\mathbf{E} + \mathbf{U} \times \mathbf{B} - \frac{m_i}{\rho_m e}\mathbf{J} \times \mathbf{B} - \frac{\hbar^2}{2e m_e}\nabla\left(\frac{\nabla^2\sqrt{\rho_m}}{\sqrt{\rho_m}}\right)\right]. \tag{6.65}$$

In (6.62), V_s is the adiabatic speed of sound of the fluid. Gauss law for magnetism can be regarded as the initial condition for Faraday's law, since $\partial(\nabla \cdot \mathbf{B})/\partial t = 0$ from (6.63). Moreover, the Hall term $\mathbf{J} \times \mathbf{B}$ at (6.65) is usually neglected.

Taking into account the equation of state, we are left with a system of 13 equations for 13 unknowns, namely, ρ_m and the components of $\mathbf{U}, \mathbf{J}, \mathbf{B}$, and $\mathbf{E}$. In comparison to classical magnetohydrodynamics, the difference of the quantum model rests on the presence of the last terms in the right-hand side of (6.61) and (6.65). These are reminiscent of the Bohm potential and can be relevant only for large density fluctuations.

In ideal magnetohydrodynamics, one assumes an infinite conductivity and neglect the Hall force in (6.65). This gives the following ideal quantum magnetohydrodynamic model,

$$\rho_m\left(\frac{\partial \mathbf{U}}{\partial t} + \mathbf{U}\cdot\nabla\mathbf{U}\right) = -\nabla P + \frac{1}{\mu_0}(\nabla\times\mathbf{B})\times\mathbf{B} + \frac{\hbar^2\rho_m}{2m_e m_i}\nabla\left(\frac{\nabla^2\sqrt{\rho_m}}{\sqrt{\rho_m}}\right), \quad (6.66)$$

$$\frac{\partial\mathbf{B}}{\partial t} = \nabla\times(\mathbf{U}\times\mathbf{B}), \qquad (6.67)$$

supplemented by the continuity equation (6.60) and the equation of state (6.62). After solving this system of seven equations for the seven unknowns $\rho_m, \mathbf{U}, \mathbf{B}$, one find the electric field from

$$\mathbf{E} = -\mathbf{U}\times\mathbf{B} + \frac{\hbar^2}{2em_e}\nabla\left(\frac{\nabla^2\sqrt{\rho_m}}{\sqrt{\rho_m}}\right). \qquad (6.68)$$

At the end, the quantum ideal magnetohydrodynamic model have the two additional terms $\sim\hbar^2$ in (6.66) and (6.68), not present in the classical case. However, (6.67) have no quantum correction. This "dynamo" equation can be used to prove that the magnetic field lines are "frozen" in the plasma, in the ideal, infinite conductivity limit. Actually, even for finite conductivity, the diffusion of magnetic field lines is described by the same diffusion equation as in classical magnetohydrodynamics. This comes from the fact that the quantum correction disappear after taking the curl of both sides of (6.65), neglecting the Hall term and assuming a constant σ as usual.

It is relevant to measure the strength of the quantum effects. For this purpose, a sensible rescaling is a good approach. We use

$$\bar\rho_m = \frac{\rho_m}{\rho_0}, \quad \bar{\mathbf{U}} = \frac{\mathbf{U}}{V_A}, \quad \bar{\mathbf{B}} = \frac{\mathbf{B}}{B_0},$$

$$\bar{\mathbf{r}} = \frac{\Omega_i\mathbf{r}}{V_A}, \quad \bar{t} = \Omega_i t, \qquad (6.69)$$

where ρ_0 and B_0 are the equilibrium mass density and magnetic field. In addition, $V_A = (B_0^2/(\mu_0\rho_0))^{1/2}$ is the Alfvén velocity, which is the natural velocity scale in magnetohydrodynamics. Moreover, $\Omega_i = eB_0/m_i$ is the ion cyclotron frequency, providing a suitable slow time scale Ω_i^{-1}. To preserve the form of the continuity equation, one then is obliged to adopt the length scale V_A/Ω_i, as shown in (6.69).

The rescaling of (6.60), (6.66) and (6.67) gives the ideal quantum magnetohydrodynamic model in nondimensional form,

$$\frac{\partial \bar{\rho}_m}{\partial t} + \nabla \cdot (\bar{\rho}_m \bar{\mathbf{U}}) = 0,$$ (6.70)

$$\bar{\rho}_m \left(\frac{\partial \bar{\mathbf{U}}}{\partial t} + \bar{\mathbf{U}} \cdot \nabla \bar{\mathbf{U}} \right) = -\frac{V_s^2}{V_A^2} \nabla \bar{\rho}_m + (\nabla \times \bar{\mathbf{B}}) \times \bar{\mathbf{B}} + \frac{H^2 \bar{\rho}_m}{2} \nabla \left(\frac{\nabla^2 (\bar{\rho}_m)^{1/2}}{(\bar{\rho}_m)^{1/2}} \right),$$ (6.71)

$$\frac{\partial \bar{\mathbf{B}}}{\partial t} = \nabla \times (\bar{\mathbf{U}} \times \bar{\mathbf{B}}),$$ (6.72)

where

$$H = \frac{\hbar \Omega_i}{\sqrt{m_e m_i} \, V_A^2}$$ (6.73)

is a dimensionless parameter measuring the relevance of quantum effects. Using MKS units, we have $H = 5.44 \times 10^{-31} n_0 / B_0$, where n_0 is the equilibrium number density. While for ordinary plasmas H is negligible, for dense astrophysical plasmas [7], with n_0 about 10^{29}–10^{34} m^{-3}, H can be significant. Hence, in dense astrophysical plasmas like the atmosphere of neutron stars or the interior of massive white dwarfs, quantum corrections to magnetohydrodynamics can be of experimental importance. However, quantum effects can be small even for large H. This happens in the cases where the density is slowly varying in space, so that the Bohm potential term in (6.71) is not of order unity.

6.4 Quantum Ideal Magnetohydrodynamics: Equilibrium Solutions

We will derive some equilibrium solutions of the quantum ideal magnetohydrodynamic model, to better understand the rôle of the quantum effects. The relevant question about the stability of these equilibria will be not addressed here.

Assuming that there is no flow ($\mathbf{U} = 0$) and that all quantities are time-independent, the ideal quantum magnetohydrodynamic model reduces to

$$\mathbf{E} = \frac{\hbar^2}{2em_e} \nabla \left(\frac{\nabla^2 \sqrt{\rho_m}}{\sqrt{\rho_m}} \right),$$ (6.74)

$$\nabla P = \frac{1}{\mu_0} (\nabla \times \mathbf{B}) \times \mathbf{B} + \frac{\hbar^2 \rho_m}{2m_e m_i} \nabla \left(\frac{\nabla^2 \sqrt{\rho_m}}{\sqrt{\rho_m}} \right).$$ (6.75)

According to (6.74), the equilibrium solutions of ideal quantum magnetohydrodynamics are not electric field free as in the classical case.

In the present analysis, we postulate a magnetic field

$$\mathbf{B} = B(r, \phi)\hat{z}, \tag{6.76}$$

in terms of cylindrical coordinates (r, ϕ, z). Gauss law is immediately satisfied. Translational invariance $(\partial/\partial z = 0)$ will be also imposed to all quantities. Moreover, straightforward algebra convert (6.75) into

$$\nabla\left(P + \frac{B^2}{2\mu_0}\right) = \frac{\hbar^2 \rho_m}{2m_e m_i} \nabla\left(\frac{\nabla^2 \sqrt{\rho_m}}{\sqrt{\rho_m}}\right). \tag{6.77}$$

Taking into account $\rho_m \simeq m_i n$ in terms of the number density n, it is advisable to restrict to a magnetic field strength such that

$$B = B(n), \tag{6.78}$$

a function of the number density only. In this case, using some equation of state $P = P(n)$, it is possible to introduce a generalized enthalpy, or effective potential $W(n)$ by means of

$$W(n) = \int^n \frac{dn'}{n'} \frac{d}{dn'}\left(P(n') + \frac{B^2(n')}{2\mu_0}\right), \tag{6.79}$$

so that

$$\nabla W = \frac{\hbar^2}{2m_e} \nabla\left(\frac{\nabla^2 \sqrt{n}}{\sqrt{n}}\right). \tag{6.80}$$

It is interesting to compare the effective potentials in (4.34) and (6.79). The latter include magnetic fields.

Defining the amplitude $n = A^2$, one has from integration of (6.80) that

$$\nabla^2 A + \frac{m_e^2 v_0^2}{\hbar^2} A = \frac{2m_e}{\hbar^2} W(A^2), \tag{6.81}$$

in terms of an integration constant v_0^2 (which can assume any sign). Notice that from (6.80) if quantum effects were absent one would be restricted to $W(n) = $ constant, a much more limited class of solutions.

Due to the quantum effects, one can find a very general class of oscillatory solutions from (6.81). To solve this equation, one needs to first stipulate the specific form of $W(n)$. If A can be found either exactly or numerically, then one can finally derive the magnetic field inverting (6.79), using the equation of state.

For a specific example, we take the simplest possible choice, namely $v_0^2 > 0$ and

$$W = 0 \Rightarrow P + \frac{B^2}{2\mu_0} = P_0, \quad P_0 = \text{cte}. \tag{6.82}$$

Hence, (6.81) reduces to the Helmholtz equation

$$\nabla^2 A + \frac{m_e^2 v_0^2}{\hbar^2} A = 0,$$ (6.83)

with general (nonsingular) solution

$$A = \sum_{v=0}^{\infty} [\alpha_v \cos(v\phi) + \beta_v \sin(v\phi)] J_v \left(\frac{m_e v_0 r}{\hbar}\right),$$ (6.84)

where the J_v are Bessel functions of the first kind and the α_v, β_v are arbitrary constants. From the constraint (6.82), one then get the magnetic field strength, provided some equation of state is specified. The oscillatory structure is evident from (6.84), unlike in the classical case. Hence, again we verify the qualitative differences between classical and quantum plasmas.

To illustrate, consider the equation of state for a zero-temperature Fermi gas in three spatial dimensions,

$$P = P_0 \left(\frac{n}{n_0}\right)^3, \quad P_0 = \frac{m_e n_0 v_{Fe}^2}{5},$$ (6.85)

where n_0 is the equilibrium number density and v_{Fe} the electron Fermi velocity. Notice that the electron Fermi energy is much higher than the ion Fermi energy, due to the smaller electron mass. Hence, the ion pressure contribution can be neglected.

Restricting to the radially symmetric solutions, or $\alpha_v = \sqrt{n_0}\delta_{v0}, \beta_v = 0$, one has from (6.82), (6.84) and (6.85)

$$\frac{n}{n_0} = \frac{A^2}{n_0} = J_0^2 \left(\frac{m_e v_0 r}{\hbar}\right),$$ (6.86)

$$\frac{P}{P_0} = \left[J_0 \left(\frac{m_e v_0 r}{\hbar}\right)\right]^6,$$ (6.87)

$$\frac{B^2}{2\mu_0 P_0} = 1 - \left[J_0 \left(\frac{m_e v_0 r}{\hbar}\right)\right]^6,$$ (6.88)

with graphics shown in Fig. 6.1. Other examples can be easily constructed. Additional translational invariant ideal quantum magnetohydrodynamic equilibria can be found in [10].

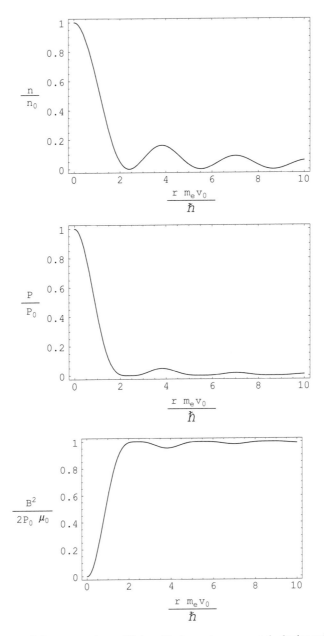

Fig. 6.1 Exact radially symmetric equilibrium ideal quantum magnetohydrodynamic solutions from (6.86)–(6.88). Density (*upper*), pressure (*mid*) and magnetic field strength (*bottom*) oscillations are shown. Nondimensional variables are used

6.5 Quantum Harris Sheet Solutions

Besides quantum magnetohydrodynamic models, we can also investigate equilibria for one-component quantum plasmas composed by an electronic fluid in a neutralizing, fixed ionic background. In this context it is natural to discuss the quantum counterpart of the Harris sheet solution [12], a celebrated classical magnetized plasma equilibrium. As will be seen, the quantum solutions exhibits an oscillatory pattern not present in the classical case. The general theory for the quantum Harris sheet solution was presented in [11].

For a one-component quantum plasma, the electromagnetic quantum fluid equations are given by (6.20) and (6.39), or

$$\frac{\partial n}{\partial t} + \nabla \cdot (n\mathbf{u}) = 0, \tag{6.89}$$

$$\frac{\partial \mathbf{u}}{\partial t} + \mathbf{u} \cdot \nabla \mathbf{u} = -\frac{1}{mn}\nabla P - \frac{e}{m}(\mathbf{E} + \mathbf{u} \times \mathbf{B}) + \frac{\hbar^2}{2m^2}\nabla \left(\frac{\nabla^2 \sqrt{n}}{\sqrt{n}} \right), \tag{6.90}$$

where $P = P(n)$ denotes the kinetic pressure. In addition, Maxwell's equations are satisfied.

Following Harris [12], we postulate purely magnetic one-dimensional stationary solutions with $\mathbf{E} = 0$ and

$$\mathbf{B} = B_y(x)\hat{y} + B_z(x)\hat{z},$$

$$n = n(x),$$

$$\mathbf{u} = u_y(x)\hat{y} + u_z(x)\hat{z},$$

$$P = P(n). \tag{6.91}$$

The continuity equation is satisfied in advance.

In terms of a vector potential $\mathbf{A}$ so that $\mathbf{B} = \nabla \times \mathbf{A}$, we choose

$$\mathbf{A} = A_y(x)\hat{y} + A_z(x)\hat{z} \Rightarrow B_y = \frac{-\mathrm{d}A_z}{\mathrm{d}x}, \quad B_z = \frac{\mathrm{d}A_y}{\mathrm{d}x}. \tag{6.92}$$

To assure charge neutrality, an appropriate immobile ionic background ionic density $n_i(x) = n(x)$ should be included, so that we don't need to worry about Poisson's equation. However, Ampère–Maxwell law $\nabla \times \mathbf{B} = \mu_0 \mathbf{J}$ with a current density $\mathbf{J} = -en\mathbf{u}$ given in terms of (6.91) gives

$$\frac{\mathrm{d}^2 A_y}{\mathrm{d}x^2} = e\mu_0 n u_y, \tag{6.93}$$

$$\frac{\mathrm{d}^2 A_z}{\mathrm{d}x^2} = e\mu_0 n u_z. \tag{6.94}$$

Moreover, the momentum transport equation yield

$$\frac{dP}{dx} = -en\left(u_y\frac{dA_y}{dx} + u_z\frac{dA_z}{dx}\right) + \frac{\hbar^2 n}{2m}\frac{d}{dx}\left(\frac{d^2\sqrt{n}/dx^2}{\sqrt{n}}\right). \tag{6.95}$$

Equations (6.93)–(6.95) form the basic system to be analyzed.

At this point, we could eliminate the velocity components from (6.93) and (6.94), substituting them in (6.95) to find

$$\frac{1}{n}\frac{d}{dx}\left(P + \frac{B^2}{2\mu_0}\right) = \frac{\hbar^2}{2m}\frac{d}{dx}\left(\frac{d^2\sqrt{n}/dx^2}{\sqrt{n}}\right), \quad B^2 = \left(\frac{dA_y}{dx}\right)^2 + \left(\frac{dA_z}{dx}\right)^2, \tag{6.96}$$

in total similarity with (6.77). Therefore, it could be straightforward to proceed as in the case of the magnetostatic equilibria of the last section, assuming $B = B(n)$, defining a generalized potential, choosing some equation of state and proceeding to the analytic solution. From this approach, we already know that (6.93)–(6.95) have spatially periodic solutions.

However, a different strategy can be followed. As proposed in [2], it is useful to restrict to the cases where (6.93) and (6.94) are derivable from a potential function $V(A_y, A_z)$, or

$$nu_y = -\frac{1}{e\mu_0}\frac{\partial V}{\partial A_y} \Rightarrow \frac{d^2 A_y}{dx^2} = -\frac{\partial V}{\partial A_y}, \tag{6.97}$$

$$nu_z = -\frac{1}{e\mu_0}\frac{\partial V}{\partial A_z} \Rightarrow \frac{d^2 A_z}{dx^2} = -\frac{\partial V}{\partial A_z}. \tag{6.98}$$

In this Hamiltonian form, some freedom is lost: A_y and A_z are not two arbitrary functions. Rather they should be found from the solution of a dynamical system involving only one input function, the potential V. However, we can now make use of the extensive knowledge on autonomous, two-dimensional Hamiltonian systems to build interesting classes of solutions. Moreover, (6.96) is reformulated in the more compact form

$$\frac{d}{dx}\left(P - \frac{V}{\mu_0}\right) = \frac{\hbar^2 n}{2m}\frac{d}{dx}\left(\frac{d^2\sqrt{n}/dx^2}{\sqrt{n}}\right), \tag{6.99}$$

using $V = -B^2/2$ apart from an irrelevant additive constant. After choosing V and solving for the vector potential components, (6.99) becomes a third-order ordinary differential equation for the density once an equation of state is plugged in.

The left-hand side of (6.96) and (6.99) is associated with the usual kinetic plus magnetic pressures balance equation of classical plasmas. The right-hand side has a quantum origin, accounting for the extra dispersion due to the Bohm potential.

It is useful to express (6.99) in terms of a variable $a = \sqrt{n}$. Taking into account the equation of state $P = P(n)$, we get

$$aa''' - a'a'' + f(a)a' + g(x) = 0, \qquad (6.100)$$

where the prime denotes differentiation with respect to x and we have introduced the quantities

$$f(a) = -\frac{4ma}{\hbar^2}\frac{dP}{dn}(n = a^2), \qquad (6.101)$$

$$g(x) = \frac{2m}{\mu_0\hbar^2}\frac{d\tilde{V}}{dx}, \quad \tilde{V}(x) = V(A_y(x), A_z(x)). \qquad (6.102)$$

The strategy to derive the solutions is now clear. Choosing a pseudo-potential $V(A_y, A_z)$ and then solving (6.97)–(6.98) for the vector potential, we determine simultaneously the magnetic field and $\tilde{V}$. Finally, we need to solve (6.100) for a specific equation of state.

To allow for a strict comparison with the classical Harris sheet solution, we should suppose a Maxwell–Boltzmann isothermal equation of state, or

$$P = n\kappa_B T, \qquad (6.103)$$

where T is the electron fluid temperature. Other equations of state (e.g., for an ultra-cold Fermi gas) could have been equally applied. However, here we are interested just on the rôle of the quantum diffraction effects represented by the right-hand side of (6.99). Hence, a classical equation of state is sufficient.

In addition, we take the following potential function,

$$V = \frac{B_\infty^2}{2}\exp\left(\frac{2A_z}{B_\infty L}\right), \qquad (6.104)$$

where L is a characteristic length and B_∞ is a reference magnetic field. The Hamiltonian system ((6.97) and (6.98)) is then

$$\frac{d^2A_y}{dx^2} = 0, \quad \frac{d^2A_z}{dx^2} = -\frac{B_\infty}{L}\exp\left(\frac{2A_z}{B_\infty L}\right). \qquad (6.105)$$

If we further take the boundary conditions $A_z(x = 0) = (dA_z/dx)(x = 0) = 0$, we get

$$A_y = A_{y0} + B_0 x, \quad A_z = -B_\infty L \ln\cosh\left(\frac{x}{L}\right), \qquad (6.106)$$

where A_{y0} and B_0 are integration constants. From the vector potential, we derive the Harris sheet magnetic field

$$B_y = B_\infty \tanh\left(\frac{x}{L}\right), \quad B_z = B_0, \qquad (6.107)$$

with an additional superimposed homogeneous magnetic field. Equation (6.107) shows a monotonic, tanh-like dependence, appropriated to situations with a smooth magnetic field polarity inversion. Classically [12] this is associated with a density bump.

If desired, the velocity field can be found from (6.93) and (6.94), which in this case becomes

$$u_y = 0, \quad u_z = \frac{B_\infty}{e\mu_0 nL} \operatorname{sech}^2\left(\frac{x}{L}\right). \tag{6.108}$$

While the magnetic field have no trace of $\hbar$, the same did not apply to the velocity field since it contain the density n. The current density, however, is purely classical.

To obtain the density n, we have to solve the third-order equation (6.100), constructed in terms of the functions $f(a)$ and $g(x)$ in (6.101) and (6.102). Using the isothermal equation of state, the potential V in (6.104) and the Harris sheet solution, we get

$$f(a) = -\frac{4m\kappa_B T}{\hbar^2} a, \tag{6.109}$$

$$g(x) = -\frac{mB_\infty^2}{\mu_0\hbar^2 L} \operatorname{sech}^2\left(\frac{x}{L}\right) \tanh\left(\frac{x}{L}\right). \tag{6.110}$$

Let us define the dimensionless variables

$$\alpha = \frac{a}{\sqrt{n_0}}, \quad X = \frac{x}{L}, \tag{6.111}$$

where n_0 is some ambient density such that

$$n_0\kappa_B T = \frac{B_\infty^2}{4\mu_0}. \tag{6.112}$$

Now (6.100) becomes

$$\alpha\frac{d^3\alpha}{dX^3} - \frac{d\alpha}{dX}\frac{d^2\alpha}{dX^2} - \frac{\alpha}{H^2}\frac{d\alpha}{dX} = \frac{1}{H^2} \operatorname{sech}^2 X \tanh X, \tag{6.113}$$

including the dimensionless parameter

$$H = \frac{\hbar}{mV_A L}, \tag{6.114}$$

where $V_A = B_\infty/(\mu_0 m n_0)^{1/2}$ is the Alfvén velocity.

The parameter H is a measure of the relevance of the quantum effects. It is the ratio of Planck's constant over 2π to the action of a particle of mass m traveling with the Alfvén velocity and confined in a length L related to the thickness of the sheet. The larger the ambient density n_0 and the smaller the characteristic length L or the characteristic magnetic field B_∞, the larger are the quantum effects. This is in accordance with the fact that quantum diffraction effects are enhanced by larger densities and smaller dimensions. The magnetic field tend to enhance the classical character of the model because the magnetic pressure $B^2/(2\mu_0)$ is additive to the kinetic pressure P. Hence, comparatively the Bohm potential term becomes less important. From this perspective, magnetized plasmas are less quantum than electrostatic plasmas. However, in the present modeling, we are disregarding a fundamental quantum contribution which is certainly decisive in magnetized plasmas: the spin [5].

To understand the rôle of the quantum terms, we may investigate (6.113) with

$$\alpha(X=0)=1, \quad \frac{d\alpha}{dX}(X=0)=0, \quad \frac{d^2\alpha}{dX^2}(X=0)=-1, \qquad (6.115)$$

which reproduces the boundary conditions for the classical Harris sheet solution, which is $\alpha = \mathrm{sech}X$. Integrating (6.113) taking into account (6.115) gives

$$\alpha\frac{d^2\alpha}{dX^2} - \left(\frac{d\alpha}{dX}\right)^2 + 1 = \frac{1}{2H^2}\left(\alpha^2 - \mathrm{sech}^2 X\right). \qquad (6.116)$$

In the ultra-quantum limit $H \gg 1$, the left-hand side of (6.116) vanishes. In this situation and using the prescribed boundary conditions, the solution would be

$$\alpha = \cos X, \quad H \gg 1. \qquad (6.117)$$

On the opposite, $H \ll 1$ case, the right-hand side of (6.116) is much bigger than the left-hand one, so that

$$\alpha = \mathrm{sech}X, \quad H \ll 1. \qquad (6.118)$$

Hence, we see a qualitative change from classical localized to quantum oscillatory solutions.

For intermediate values of the quantum parameter, (6.116) can be numerically investigated with initial condition $\alpha(0)=1, \alpha'(0)=0$. For instance, Fig. 6.2 show the results for $H = 1$ and $H = 5$. For larger quantum parameter, we observe a slow amplitude decay, indicating the oscillatory structure culminating with periodic solutions in the extreme quantum limit.

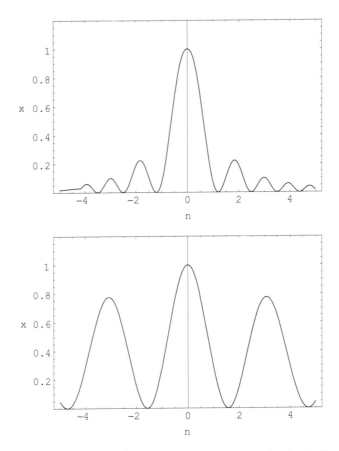

Fig. 6.2 Density oscillations $n = n_0 \alpha^2$ from (6.116). Parameters: $n_0 = L = 1$. *Top*: $H = 1$. *Bottom*: $H = 5$

Problems

6.1. Fill up the missing points and derive the electromagnetic Wigner equation (6.3).

6.2. Show that the electromagnetic Wigner equation (6.3) is not gauge invariant.

6.3. Use the change of variables (6.13) and derive Vlasov's equation as the classical limit of the electromagnetic Wigner equation (6.3).

6.4. Demonstrate the identity (6.7).

6.5. Derive the hydromagnetic quantum model (6.20) and (6.21).

6.6. Verify (6.33).

6.7. Defining an average osmotic velocity

$$\mathbf{u}^{\circ} = \sum_{\alpha=1}^{M} \tilde{p}_{\alpha} \mathbf{u}_{\alpha}^{\circ},$$

with modified statistical weights in (6.34), show that it satisfies the simple relation

$$\mathbf{u}^{\circ} = \frac{\hbar}{2m} \frac{\nabla n}{n}.$$

6.8. Relate the average osmotic velocity to the Bohm potential term.

6.9. Derive the quantum magnetohydrodynamic equations (6.55) and (6.56) from the two-fluid quantum plasma model, following the steps indicated in the text.

References

1. Arnold, A. and Steinrück, H: The 'electromagnetic' Wigner equation for an electron with spin. J. Appl. Math. Phys. **40**, 793–815 (1989)
2. Attico, N. and Pegoraro, F.: Periodic equilibria of the VlasovMaxwell system. Phys. Plasmas **6**, 767–771 (1999)
3. Bittencourt, J. A.: Fundamentals of Plasma Physics. National Institute for Space Research, São José dos Campos (1995)
4. Bohm, D. and Hiley, B. J.: The Undivided Universe: an Ontological Interpretation of Quantum Theory. Routledge, London (1993)
5. Brodin, G. and Marklund, M.: Spin magnetohydrodynamics. New J. Phys. **9**, 277–289 (2007)
6. Carruthers, P., Zachariasen, F.: Quantum collision theory with phase-space distributions. Rev. Mod. Phys. **55**, 245–285 (1983)
7. Chabrier, G., Douchin, F. and Potekhin, A. F.: Dense astrophysical plasmas. J. Phys.: Cond. Matter **14**, 9133–9139 (2002)
8. de Groot, S. R. and Suttorp, L. G.: Foundations of Electrodynamics. North-Holland, Amsterdam (1972)
9. Gasser, I., Lin, C. and Markowich, P. A.: A review of dispersive limits of (non)linear Schrödinger-type equations. Taiwan. J. Math. **4**, 501–529 (2000)
10. Haas, F.: A magnetohydrodynamic model for quantum plasmas. Phys. Plasmas **12**, 062117–062126 (2005)
11. Haas, F.: Harris sheet solution for magnetized quantum plasmas. Europhys. Lett. **77**, 45004–45009 (2007)
12. Harris, E. G.: On a plasma sheath separating regions of oppositely directed magnetic field. Nuovo Cimento **23** 115–121 (1962)
13. Masmoudi, N. and Mauser, N. J.: The selfconsistent Pauli equation. Monatshefte für Mathematik, **132**, 19–24 (2001)
14. Materdey, T. B. and Seyler, C. E.: The quantum Wigner function in a magnetic field. Int. J. Mod. Phys. B **17**, 4555–4592 (2003)
15. Nicholson, D. R.: Introduction to Plasma Theory. John Wiley, New York (1983)
16. Saikin, S.: A drift-diffusion model for spin-polarized transport in a two-dimensional non-degenerate electron gas controlled by spinorbit interaction. J. Phys.: Condens. Matter **16**, 5071–5081 (2004)

Chapter 7
The One-Dimensional Quantum Zakharov System

Abstract In classical plasma, the Zakharov system provides the fluid modeling of the interaction between Langmuir and ion-acoustic waves. The derivation of the corresponding quantum Zakharov system from the quantum two-fluid system is shown in detail. Applications are shown for quantum parametric and four-wave instabilities. A Lagrangian approach is used for the semiclassical adiabatic regime. This case reduces to a generalized nonlinear Schrödinger equation for the envelope electric field. A time-dependent variational formalism is applied in the analysis of the internal oscillations of quantum Langmuir solitons, in the general, nonadiabatic and nonsemiclassical case.

7.1 Quantum Zakharov Equations in One Spatial Dimension

In one spatial dimension and using rescaled variables, the classical Zakharov system can be expressed as

$$i\frac{\partial \tilde{E}}{\partial t} + \frac{\partial^2 \tilde{E}}{\partial x^2} = n\tilde{E}, \tag{7.1}$$

$$\frac{\partial^2 n}{\partial t^2} - \frac{\partial^2 n}{\partial x^2} = \frac{\partial^2 |\tilde{E}|^2}{\partial x^2}, \tag{7.2}$$

where $\tilde{E}$ is the slowly varying envelope of the high frequency electric and n the plasma density perturbation from its equilibrium value. Since its formulation by Zakharov [19] the system (7.1) and (7.2) has been the subject of several theoretical and experimental works, as can be verified in reviews [18]. As will become clear from the examples below, interacting Langmuir (high frequency) and ion-acoustic (low frequency) waves are supported by the usual and quantum Zakharov systems. A detailed account on the nonlinear energy transfer between these fundamental plasma oscillations justify the Zakharov's system popularity.

F. Haas, *Quantum Plasmas: An Hydrodynamic Approach*, Springer Series on Atomic, Optical, and Plasma Physics 65, DOI 10.1007/978-1-4419-8201-8_7,
© Springer Science+Business Media, LLC 2011

The derivation of the model follows precisely from a two-time scales analysis, made possible due to the presence of high and low frequency oscillations. The starting point are the fluid equations for an electron–ion plasma, so that kinetic effects such as Landau damping are not included. Hence, only wavenumbers $k \ll k_D$ are admitted, where $k_D = 2\pi/\lambda_D$ is the Debye wavenumber, with λ_D being the Debye length. In the degenerate case, we need to consider the corresponding Thomas–Fermi length λ_F and Fermi wavenumber $k_F = 2\pi/\lambda_F$. A kinetic treatment of the interaction between quantum Langmuir and ion-acoustic waves can be found in [13]. In addition, "almost quasi" neutrality is considered to be valid, in the sense that departures from quasi-neutrality occur only for the high frequency part of the electron fluid density. Finally, a weak turbulence or weak correlation hypothesis should be assumed, as detailed in the continuation. A general discussion about the derivation of the (classical) Zakharov equations can be found in [18]. The quantum Zakharov system was introduced in [5].

Consider an one-dimensional two-species electrostatic quantum plasma decribed by

$$\frac{\partial n_e}{\partial t} + \frac{\partial(n_e u_e)}{\partial x} = 0, \tag{7.3}$$

$$\frac{\partial n_i}{\partial t} + \frac{\partial(n_i u_i)}{\partial x} = 0, \tag{7.4}$$

$$\frac{\partial u_e}{\partial t} + u_e \frac{\partial u_e}{\partial x} = -\frac{1}{m_e n_e}\frac{\partial P_e}{\partial x} - \frac{eE}{m_e} + \frac{\hbar^2}{2m_e^2}\frac{\partial}{\partial x}\left(\frac{\partial^2\sqrt{n_e}/\partial x^2}{\sqrt{n_e}}\right), \tag{7.5}$$

$$\frac{\partial u_i}{\partial t} + u_i \frac{\partial u_i}{\partial x} = \frac{e}{m_i}E, \tag{7.6}$$

$$\frac{\partial E}{\partial x} = \frac{e}{\varepsilon_0}(n_i - n_e), \tag{7.7}$$

where $n_{e,i}$ are the electron or ion fluid densities, $u_{e,i}$ the electron or ion fluid velocities, E the electric field, P_e the electron fluid pressure, $m_{e,i}$ the electron or ion masses. Ions are single charged with a charge $+e$. Finally, ε_0 and $\hbar = h/(2\pi)$ are, respectively, the vacuum permittivity and Planck's constant over 2π. Hereafter, assume $m_i \gg m_e$, a necessary assumption to have a two-time scales separation due to the slow and rapid time scales associated with ions and electrons, respectively. Therefore, the following treatment is not applicable, for example, to electron–positron plasmas. Quantum effects are not present in the ion momentum equation, also due to the larger ion mass. Indeed the Bohm potential term is proportional to the inverse squared mass. Finally, for simplicity ions are assumed to be cold (pressureless). Hence, we have the same initial model of Chap. 5 on the quantum ion-acoustic wave.

The electron fluid equation of state is not very relevant because after some rescalings and series expansions the specific form of P_e can be hardly detected.

For definiteness, we consider the zero-temperature one-dimensional Fermi gas equation of state

$$P_e = \frac{m_e n_0 v_{Fe}^2}{3} \left(\frac{n_e}{n_0}\right)^3, \tag{7.8}$$

where n_0 is the equilibrium electron fluid density and v_{Fe} the electron Fermi velocity.

Quantum ion-acoustic waves were obtained from (7.3)–(7.7) in Chap. 5, after disregarding the electron's inertia. Let us briefly discuss the quantum Langmuir waves also supported by the same model. Linearization of the electron equations (7.3), (7.5) and (7.7) around

$$n_e = n_i = n_0, \quad u_e = 0, \quad E = 0 \tag{7.9}$$

and Fourier analyzing yield

$$\omega^2 = \omega_{pe}^2 + 3 v_{Fe}^2 k^2 + \frac{\hbar^2}{4 m_e^2} k^4, \tag{7.10}$$

where ω is the wave frequency, k is the wavenumber and $\omega_{pe} = (n_0 e^2 / m_e \varepsilon_0)^{1/2}$ the electron plasma frequency. Taking the formal classical limit $\hbar \to 0$ gives the classical Langmuir waves dispersion relation [14]. There is no damping or instability in the corresponding quantum Langmuir waves described by (7.10). Only in a kinetic treatment or some extended fluid approach, these features could have been included.

We now proceed to the derivation of the quantum Zakharov system following as closely as possible the method applied to the classical fluid equations [14, 18, 19], the only noticeable difference being the presence of the quantum potential in the electron momentum equation. First, decompose the fluid variables into fast and slow oscillatory parts, identified by the subscripts f and s, respectively,

$$n_e(x,t) = n_0 + n_s(x,t) + n_f(x,t), \tag{7.11}$$

$$n_i(x,t) = n_0 + n_s(x,t), \tag{7.12}$$

$$u_e(x,t) = u_s(x,t) + u_f(x,t), \tag{7.13}$$

$$u_i(x,t) = u_s(x,t), \tag{7.14}$$

$$E(x,t) = E_s(x,t) + E_f(x,t). \tag{7.15}$$

We are considering that the fast quantities have zero average, while the slow quantities does not significantly change over an oscillation period. This is very much in the spirit of a two-time scales analysis [3]. The high frequency components of the ion fluid density and velocity were disregarded in view of $m_i \gg m_e$. In addition, quasi-neutrality ($n_i \approx n_e$ and $u_i \approx u_e$) is almost exact, except from the high frequency oscillatory electronic contributions.

For completeness, it is convenient to review the whole classical derivation in detail, now including the Bohm potential term. Taking into account the ion continuity and force equations (7.4) and (7.6), we have from (7.11)–(7.15) and the electron equations (7.3) and (7.5) that

$$\frac{\partial n_f}{\partial t} + n_0 \frac{\partial u_f}{\partial x} + \frac{\partial (n_s u_f)}{\partial x} + \frac{\partial (n_f u_s)}{\partial x} + \frac{\partial (n_f u_f)}{\partial x} = 0, \tag{7.16}$$

$$\frac{\partial u_f}{\partial t} + u_s \frac{\partial u_f}{\partial x} + u_f \frac{\partial u_s}{\partial x} + u_f \frac{\partial u_f}{\partial x} = -\frac{e}{m_e}(E_s + E_f) - 3\frac{v_{Fe}^2}{n_0} \frac{\partial (n_s + n_f)}{\partial x}$$
$$+ \frac{\hbar^2}{4m_e^2 n_0} \frac{\partial^3 (n_s + n_f)}{\partial x^3}. \tag{7.17}$$

The electron density was linearized around n_0. Here, it becomes clear that basically the same results would follow from another equation of state, due to the neglect of higher-order density fluctuation effects. Finally, a term was disregarded in (7.17) in view of $m_i \gg m_e$.

Differentiating Poisson's equation with respect to t, using the continuity equations and integrating with respect to x we get

$$\frac{\partial E}{\partial t} = \frac{e}{\varepsilon_0}(n_e u_e - n_i u_i), \tag{7.18}$$

which is Ampère's law for a zero magnetic field. Subtracting from (7.18) the average of itself, the result is

$$\frac{\partial E_f}{\partial t} = \frac{e}{\varepsilon_0}(n_0 + n_s)u_f + \frac{e}{\varepsilon_0}n_f u_s + \frac{e}{\varepsilon_0}[n_f u_f - \langle n_f u_f \rangle], \tag{7.19}$$

where $\langle \rangle$ denotes time average over a period.

The third term in the right-hand side of (7.19) can be ignored in view of the following argument. Linearization of both the ion continuity equation and (7.16) yields the rough estimates

$$\omega_s n_s \sim n_0 k_s u_s, \qquad \omega_{pe} n_f \sim n_0 k u_f, \tag{7.20}$$

where ω_s^{-1}, k_s^{-1} are the time and length scales of the slowly varying functions, and ω_{pe}^{-1}, k^{-1} the time and length scales of the fast quantities. Therefore, a comparison between the second and third terms in (7.19) yield

$$\frac{n_f u_s}{n_s u_f} \sim \frac{k}{k_s} \frac{\omega_s}{\omega_{pe}} \ll 1, \tag{7.21}$$

since $\omega_s \ll \omega_{pe}$, neglecting the unlikely situation when $k \gg k_s$. Moreover, the contribution inside brackets in (7.19) can be also discarded from a weak turbulence condition [18] which allows to ignore correlations. Hence, (7.19) simplifies to

$$\frac{\partial E_f}{\partial t} = \frac{e}{\varepsilon_0}(n_0 + n_s)u_f. \tag{7.22}$$

Subtracting from (7.17) the average of itself, we get

$$\frac{\partial u_{\mathrm{f}}}{\partial t} = -\frac{e}{m_{\mathrm{e}}}E_{\mathrm{f}} - 3\frac{v_{\mathrm{Fe}}^2}{n_0}\frac{\partial n_{\mathrm{f}}}{\partial x} + \frac{\hbar^2}{4m_{\mathrm{e}}^2 n_0}\frac{\partial^3 n_{\mathrm{f}}}{\partial x^3}, \tag{7.23}$$

using once again the weak turbulence hypothesis which allows to ignore convective and correlation terms.

From the fast component of the Poisson equation,

$$\frac{\partial E_{\mathrm{f}}}{\partial x} = -\frac{e n_{\mathrm{f}}}{\varepsilon_0}. \tag{7.24}$$

Using (7.22)–(7.23), we obtain

$$\left(1 - \frac{n_{\mathrm{s}}}{n_0}\right)\frac{\partial^2 E_{\mathrm{f}}}{\partial t^2} + \omega_{\mathrm{pe}}^2 E_{\mathrm{f}} = 3v_{\mathrm{Fe}}^2\frac{\partial^2 E_{\mathrm{f}}}{\partial x^2} - \frac{\hbar^2}{4m_{\mathrm{e}}^2}\frac{\partial^4 E_{\mathrm{f}}}{\partial x^4}, \tag{7.25}$$

where again only the leading density fluctuation term was retained. The underlying strategy of keeping only the first-order density perturbations has consequences on the form of the quantum corrections, since the Bohm potential is strongly dependent on the modulations of n_{e}. However, extending the theory to higher-order terms seems to be a challenge.

The high frequency part of the electric field can be decomposed as

$$E_{\mathrm{f}}(x,t) = \frac{1}{2}\tilde{E}(x,t)\mathrm{e}^{-\mathrm{i}\omega_{\mathrm{pe}}t} + \mathrm{c.c.}, \tag{7.26}$$

where $\tilde{E}(x,t)$ is an slowly varying "envelope" and c.c. refer to complex conjugate. Hence, Langmuir oscillations are incorporated already in the exponential since they occur approximately at the electron plasma frequency. Inserting (7.26) into (7.25) yield

$$i\frac{\partial \tilde{E}}{\partial t} + \frac{3}{2}\frac{v_{\mathrm{Fe}}^2}{\omega_{\mathrm{pe}}}\frac{\partial^2 \tilde{E}}{\partial x^2} - \frac{\hbar^2}{8m_{\mathrm{e}}^2\omega_{\mathrm{pe}}}\frac{\partial^4 \tilde{E}}{\partial x^4} = \frac{\omega_{\mathrm{pe}}}{2}\frac{n_{\mathrm{s}}}{n_0}\tilde{E}. \tag{7.27}$$

where the second-order time-derivative of the envelope field has been neglected since

$$\left|\frac{\partial^2 \tilde{E}}{\partial t^2}\right| \ll \left|\omega_{\mathrm{pe}}\frac{\partial \tilde{E}}{\partial t}\right|. \tag{7.28}$$

Equation (7.27) describes the time evolution of the slowly varying amplitude $\tilde{E}$.

Equation (7.27) contain the slow part n_{s} of the plasma density perturbation. To study the dynamics of n_{s}, consider the low frequency parts of the electron continuity and ion force equations,

$$\frac{\partial n_{\mathrm{s}}}{\partial t} + n_0\frac{\partial u_{\mathrm{s}}}{\partial x} = 0, \tag{7.29}$$

$$\frac{\partial u_s}{\partial t} - \frac{e}{m_i} E_s = 0. \tag{7.30}$$

Convective terms were once more disregarded. In the same spirit, the slow part of the electron momentum equation imply

$$\frac{\partial u_s}{\partial t} + \frac{e}{m_e} E_s + 3 \frac{v_{Fe}^2}{n_0} \frac{\partial n_s}{\partial x} - \frac{\hbar^2}{4m_e^2 n_0} \frac{\partial^3 n_s}{\partial x^3} + \left\langle u_f \frac{\partial u_f}{\partial x} \right\rangle = 0. \tag{7.31}$$

The last term above can be treated using the estimate $u_f \sim -eE_f/(m_e \omega_{pe})$, writing the so-called ponderomotive force [18] according to

$$m_e \left\langle u_f \frac{\partial u_f}{\partial x} \right\rangle = \frac{e^2}{2m_e \omega_{pe}^2} \frac{\partial}{\partial x} \langle E_f^2 \rangle = \frac{e^2}{4m_e \omega_e^2} \frac{\partial |\tilde{E}|^2}{\partial x}. \tag{7.32}$$

Finally, from (7.31) we get

$$\frac{\partial u_s}{\partial t} + \frac{e}{m_e} E_s + 3 \frac{v_{Fe}^2}{n_0} \frac{\partial n_s}{\partial x} - \frac{\hbar^2}{4m_e^2 n_0} \frac{\partial^3 n_s}{\partial x^3} + \frac{e^2}{4m_e^2 \omega_{pe}^2} \frac{\partial |\tilde{E}|^2}{\partial x} = 0. \tag{7.33}$$

Eliminating u_s and E_s from (7.29), (7.30) and (7.33) and using $m_e/m_i \ll 1$, the result is

$$\frac{\partial^2 n_s}{\partial t^2} - 3c_s^2 \frac{\partial^2 n_s}{\partial x^2} + \frac{\hbar^2}{4m_i m_e} \frac{\partial^4 n_s}{\partial x^4} = \frac{\varepsilon_0}{4m_i} \frac{\partial^2 |\tilde{E}|^2}{\partial x^2}, \tag{7.34}$$

where $c_s = (\kappa_B T_{Fe}/m_i)^{1/2}$ is the quantum ion-acoustic velocity. Equations (7.27) and (7.34) form a coupled system for the envelope electric field and the slow component of the plasma density fluctuation. It is the quantum analog of the classical Zakharov system, which would be obtained if $\hbar \equiv 0$. Hence, it is natural to refer to (7.27) and (7.34) as the quantum Zakharov equations [5].

It is convenient to clean (7.27) and (7.34) from some weird factors using the normalized quantities

$$\bar{x} = 2\sqrt{\frac{m_e}{3m_i}} \frac{x}{\lambda_F}, \quad \bar{t} = 2\frac{m_e}{m_i} \omega_{pe} t, \tag{7.35}$$

$$\bar{n} = \frac{1}{4} \frac{m_i}{m_e} \frac{n_s}{n_0}, \quad \bar{E} = \frac{e\tilde{E}}{4\sqrt{3} m_e v_{Fe} \omega_{pi}}, \tag{7.36}$$

where $\lambda_F = v_{Fe}/\omega_{pe}$ is the electron Thomas–Fermi length and $\omega_{pi} = (n_0 e^2/m_i \varepsilon_0)^{1/2}$ is the ion plasma frequency. It is also convenient to introduce the dimensionless quantum parameter

$$H = \frac{\hbar \omega_{pi}}{3\kappa_B T_{Fe}}, \tag{7.37}$$

with $\kappa_B T_{Fe} = m_e v_{Fe}^2$. Dropping bars, the final system reads

$$i\frac{\partial E}{\partial t} + \frac{\partial^2 E}{\partial x^2} - H^2 \frac{\partial^4 E}{\partial x^4} = nE, \tag{7.38}$$

$$\frac{\partial^2 n}{\partial t^2} - \frac{\partial^2 n}{\partial x^2} + H^2 \frac{\partial^4 n}{\partial x^4} = \frac{\partial^2 |E|^2}{\partial x^2}. \tag{7.39}$$

The quantum parameter H in (7.37) is the ratio between the ion plasmon and electron thermal energies. The unavoidable presence of the ion parameter ω_{pi} in H is due to the existence of the ion-acoustic collective mode.

One should have in mind the numerous approximations in the derivation of the quantum Zakharov equations, in particular the nonrelativistic and weak turbulence (or weak coupling) assumptions. Nevertheless, we may expect that some of the gross features of the quantum effects in the nonlinear interaction of Langmuir and ion-sound waves in dense plasmas are still captured by (7.38) and (7.39), in a first approximation at least.

Instead of paying attention to the quantity H only, a more legitimate estimate of the quantum effects comes from the ratio of the second and third terms of the left-hand side of the dimensional (7.27) and (7.34). Assuming $\partial/\partial x \sim k$ gives

$$\frac{\left(\frac{\hbar^2}{m_e^2 \omega_e} \frac{\partial^4 \bar{E}}{\partial x^4}\right)}{\left(\frac{v_{Fe}^2}{\omega_{pe}} \frac{\partial^2 \bar{E}}{\partial x^2}\right)} \sim \frac{\left(\frac{\hbar^2}{m_e m_i} \frac{\partial^4 n_s}{\partial x^4}\right)}{\left(c_s^2 \frac{\partial^2 n_s}{\partial x^2}\right)} \sim \left(\frac{\hbar k}{m_e v_{Fe}}\right)^2 < \left(\frac{\hbar \omega_{pe}}{\kappa_B T_{Fe}}\right)^2, \tag{7.40}$$

for the relative strength of the quantum terms, the last inequality arising from the high frequency hypothesis $\omega_{pe} > k v_{Fe}$ which is also needed to avoid kinetic effects. From (7.40), we have that the quantum diffraction effects are comparable to the Fermi statistics effects provided the electron plasmon energy is of the same order of the Fermi energy.

In the following model, (7.38) and (7.39) is applied to two well-known parametric instabilities from classical plasma physics, namely the decay and four wave instabilities.

7.2 Parametric Instabilities

7.2.1 Decay Instability

Consider the *Ansatz*

$$E = E_0 e^{i(k_0 x - \omega_0 t)} + E_1(t) e^{i(k_1 x - \omega_1 t)}, \tag{7.41}$$

$$n = n_1(t) \cos(Kx - \Omega t), \tag{7.42}$$

for the quantum Zakharov equations (7.38) and (7.39), where $E_0, k_{0,1}, \omega_{0,1}, K, \Omega$ are constants and $E_1(t)$ and $n_1(t)$ are first-order quantities. This *Ansatz* mimics the classical treatment of the decay instability [18]. Here,

$$\omega_0 = k_0^2 + H^2 k_0^4, \tag{7.43}$$

$$\omega_1 = k_1^2 + H^2 k_1^4, \tag{7.44}$$

$$\Omega^2 = K^2 + H^2 K^4. \tag{7.45}$$

Equations (7.43) and (7.44) correspond to Langmuir waves, where the electron plasma frequency contributions did not appear due to the decomposition (7.26) which produces a slowly varying envelope field. On the other hand, (7.45) is associated with quantum ion-acoustic waves [10]. The quantum Zakharov system admit the plane wave solution (7.41) and (7.42) with $E_1 = n_1 = 0$, thanks to the dispersion relation (7.43). The question now is about the stability of such a solution, for increasing amplitude E_0 playing the rôle of control parameter.

Due to momentum and energy conservation [18], the matching conditions

$$k_0 = k_1 + K, \quad \omega_0 = \omega_1 + \Omega, \tag{7.46}$$

hold. Therefore, we have the decay of a quantum Langmuir wave, with dispersion relation (7.43), into other quantum Langmuir wave, with dispersion relation (7.44), and a quantum ion-acoustic wave, with dispersion relation (7.45).

Keeping only the first-order quantities in (7.38) and (7.39), we obtain

$$i\dot{E}_1 e_1 = \frac{E_0 n_1}{2}(e_+ + e_-)e_0, \tag{7.47}$$

$$\left(\frac{\ddot{n}_1}{2} - i\Omega \dot{n}_1 + K^2 E_0 E_1^*\right) e_+ + \left(\frac{\ddot{n}_1}{2} + i\Omega \dot{n}_1 + K^2 E_0^* E_1\right) e_- = 0, \tag{7.48}$$

with the notation

$$e_{0,1} = \exp[i(k_{0,1}x - \omega_{0,1}t)], \quad e_\pm = \exp[\pm i(Kx - \Omega t)]. \tag{7.49}$$

Taking into account the resonance condition (7.46), (7.47) read

$$i\dot{E}_1 = \frac{E_0 n_1}{2}\left(1 + e^{2i(Kx - \Omega t)}\right), \tag{7.50}$$

so that, apart from a zero-average oscillatory part, we have

$$n_1 = \frac{2i}{E_0}\dot{E}_1. \tag{7.51}$$

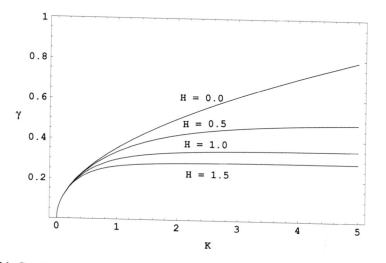

Fig. 7.1 Growth rate of the decay instability for $E_0 = 0.5$ and $0 \leq H \leq 1.5$, showing the saturation effect for $H > 0$

Similarly, (7.48) gives

$$\frac{\ddot{n}_1}{2} + i\Omega \dot{n}_1 + K^2 E_0^* E_1 = 0. \tag{7.52}$$

Assuming $\dot{E}_1 = i\omega E_1$, from (7.51) and (7.52), we get

$$\omega^3 + 2\Omega\omega^2 + K^2|E_0|^2 = 0, \tag{7.53}$$

which exactly reproduces the classical decay instability dispersion relation [18]. Hence, the conclusions valid for the classical case can be immediately extended to the quantum case. In particular, when $\Omega \gg |\omega|$, so that the cubic term can be disregarded in (7.53), defining $\omega = i\gamma$ we deduce the growth rate

$$\gamma = \frac{K|E_0|}{\sqrt{2\Omega}}. \tag{7.54}$$

With more generality, the cubic equation (7.53) can be shown to have a positive discriminant, so that it always admit one real and two complex conjugate solutions [8]. Hence, only one solution is associated with a (decay) instability, with unbounded growth of both perturbations E_1 and n_1. However, in spite of the formal similarity, the quantum case is different from the classical one due to the quantum corrections in the dispersion relations (7.43)–(7.45) for the quantum Langmuir and ion-acoustic modes.

The distinct quantum Langmuir dispersion relation produces a saturation effect not present in the classical case (see Fig. 7.1). Indeed, using (7.45) we can rewrite (7.54) as

$$\gamma = \frac{\sqrt{K}|E_0|}{\sqrt{2}(1 + H^2 K^2)^{1/4}}, \tag{7.55}$$

admitting a maximum value $\gamma_{max} = |E_0|/\sqrt{2H}$ for large K. This is to be compared with the formal classical case ($H = 0$) where γ grows with no bound as K increases. The quantum diffraction effects, therefore, did not stabilize the decay instability, but at least impose a maximum growth rate.

Notice that neglecting the first term in (7.53) is valid for large wavenumbers in both the classical and quantum situations. Indeed, for large K we have $\Omega \sim K, |\omega| \sim K^{1/2}$ in the classical case and $\Omega \sim K^2, |\omega| \sim 1$ in the quantum case, so that the first term in (7.53) becomes more and more negligible as K increases.

However, the wavenumber can not increase indefinitely otherwise the fluid model becomes incorrect. Taking into account the rescaling (7.35) and the inequality (7.40), we have

$$KH = \sqrt{\frac{3m_i}{m_e}} \frac{k\lambda_{Fe}}{2} \frac{\hbar\omega_{pi}}{3\kappa_B T_{Fe}} < \frac{\hbar\omega_{pe}}{2\sqrt{3}\kappa_B T_{Fe}}. \qquad (7.56)$$

The right-hand side of (7.56) shows an upper bound on the quantum effects strength, basically given by the ratio between the electron plasmon and Fermi energies, as expected.

7.2.2 Four-Wave Instability

Let us consider the energy transfer from one single finite-amplitude quantum Langmuir wave and two other quantum Langmuir waves and one quantum ion-acoustic wave. In comparison to the previous problem, now one extra quantum Langmuir wave is included. The appropriate model can be written [14] as

$$E(x,t) = E_0 \exp(-i\omega_0 t + ik_0 x) + E_+ \exp[-i(\omega_0 + \omega)t + i(k_0 + k)x]$$
$$+ E_- \exp[-i(\omega_0 - \omega^*)t + i(k_0 - k)x], \qquad (7.57)$$

$$n(x,t) = \tilde{n}\exp(-i\omega t + ikx) + c.c, \qquad (7.58)$$

where E_0 is a zeroth-order and $E_{\pm}$ and $\tilde{n}$ are first-order amplitudes, ω_0, k_0 and k are real constants and ω is a complex constant. To have an unperturbed solution $E(x,t) = E_0 \exp(-i\omega_0 t + ik_0 x), n(x,t) = 0$ satisfying (7.38) and (7.39), necessarily

$$\omega_0 = k_0^2 + H^2 k_0^4. \qquad (7.59)$$

which correspond to a quantum Langmuir mode after absorbing the electron plasma frequency trough the definition (7.26).

Inserting (7.57) and (7.58) into (7.38) for the time-evolution of the envelope electric field, linearizing and separating the parts proportional to $\exp[-i(\omega_0 + \omega)t + i(k_0 + k)x]$ and $\exp[-i(\omega_0 - \omega^*)t + i(k_0 - k)x]$ gives

$$(\omega_0 + \omega)E_+ - (k_0 + k)^2 E_+ = \tilde{n}E_0 + H^2(k_0 + k)^4 E_+, \tag{7.60}$$

$$(\omega_0 - \omega^*)E_- - (k_0 - k)^2 E_- = \tilde{n}^* E_0 + H^2(k_0 - k)^4 E_-. \tag{7.61}$$

In a similar way, from (7.39) for the time-evolution of the density perturbation we find

$$(\omega^2 - k^2 - H^2 k^4)\tilde{n} = k^2(E_0^* E_+ + E_0 E_-^*) \tag{7.62}$$

plus

$$((\omega^*)^2 - k^2 - H^2 k^4)\tilde{n}^* = k^2(E_0 E_+^* + E_0^* E_-). \tag{7.63}$$

In passing, we observe that (7.62) and (7.63) are compatible if and only if $\omega^2 = (\omega^*)^2$. Hence, ω should be either purely real or purely imaginary. Assuming $E_0, E_\pm$ to be real, we have from (7.60)–(7.63) an homogeneous system of four equations for the four quantities $E_\pm, \tilde{n}$ and $\tilde{n}^*$. Nontrivial solutions exist provided the dispersion relation

$$D_s D_1 D_2 = k^2 E_0^2 (D_1 + D_2), \tag{7.64}$$

is satisfied, where

$$D_s = \omega^2 - k^2 - H^2 k^4, \tag{7.65}$$

$$D_1 = \omega + \omega_0 - (k + k_0)^2 - H^2(k + k_0)^4, \tag{7.66}$$

$$D_2 = -\omega + \omega_0 - (k - k_0)^2 - H^2(k - k_0)^4. \tag{7.67}$$

In the limit $H \to 0$ the classical dispersion relation for the four-wave interaction [14] is recovered. From (7.64), we see that in the absence of the pump amplitude E_0 we have one quantum ion-acoustic ($D_s = 0$) and two quantum Langmuir ($D_{1,2} = 0$) modes, without any instability therefore.

The dispersion relation (7.64) is a fourth-order polynomial equation for ω. Fortunately, assuming $\omega_0 = k_0 = 0$ and $\omega = i\gamma$ it moves to a quadratic equation for γ^2,

$$[\gamma^2 + k^2 + H^2 k^4][\gamma^2 + (k^2 + H^2 k^4)^2] = 2k^2 E_0^2(k^2 + H^2 k^4). \tag{7.68}$$

Solving for γ^2 two roots are found,

$$\gamma^2 = -\frac{1}{2}(k^2 + H^2 k^4)(1 + k^2 + H^2 k^4)$$

$$\pm \frac{1}{2}(k^2 + H^2 k^4)^{1/2}[(k^2 + H^2 k^4)(1 - k^2 - H^2 k^4)^2 + 8k^2 E_0^2]^{1/2}. \tag{7.69}$$

One root is a stable mode ($\gamma^2 < 0$). The other one is positive for sufficiently high pump amplitude,

$$E_0^2 > \frac{k^2}{2}(1 + H^2 k^2)^2. \tag{7.70}$$

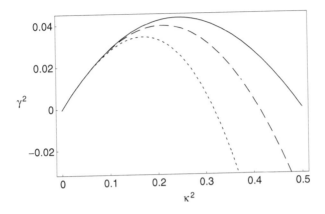

Fig. 7.2 γ^2 as a function of k^2 for the potentially unstable mode of the dispersion relation (7.69) for the four-wave instability according to [5]. Parameters: $E_0 = 0.5$, $H = 0$ (*full line*), $H = 0.5$ (*dashed line*) and $H = 0.9$ (*dotted line*)

If (7.70) is satisfied, we have a purely growing instability. In all cases, we have $\omega^2 = (\omega^*)^2$ so that (7.62) and (7.63) are consistent.

When $H \to 0$, (7.70) recovers the classical instability condition [14] for the four-wave interaction. However, for larger quantum effects it becomes increasingly difficult to satisfy the inequality (7.70). There is even suppression of the four-wave instability whenever

$$H^2 \geq \frac{\sqrt{2}E_0 - k}{k^3}. \tag{7.71}$$

Hence, we have one more example of the stabilizing rôle of the quantum diffraction effects. In this case, there is a less efficient energy transfer from the original Langmuir wave to the additional Langmuir and ion-acoustic modes, due to the Bohm potential terms in the quantum Zakharov system.

Let us study in more detail the potentially unstable mode associated with the positive root in (7.69). In Fig. 7.2, we show this γ^2 as a function of k^2 for $H = 0$, $H = 0.5$ and $H = 0.9$, always with $E_0 = 0.5$. The instability region ($\gamma^2 > 0$) in k-space becomes narrower for bigger H, and the maximum γ^2 becomes smaller for larger quantum effects. The wavenumber for maximum growth rate, k_{max} can be approximately calculated by expanding (7.69) up to $O(k^6)$, which is reasonable in view of the long wavelength approximation. Then, $d(\gamma^2)/d(k^2) = 0$ gives a quadratic equation for k^2, which can be solved in closed form. The cumbersome resulting expression will be omitted here, but the result of the expansion procedure is shown in Fig. 7.3, where k_{max}^2 for maximum growth rate is shown as a function of H for a fixed $E_0 = 0.5$. Inserting this k_{max}^2 in the unstable mode in (7.69) an awkward expression for the maximum growth rate as a function of H is derived. Figure 7.4 shows how an increasing quantum parameter produces a smaller maximum growth rate, for the same particular value $E_0 = 0.5$.

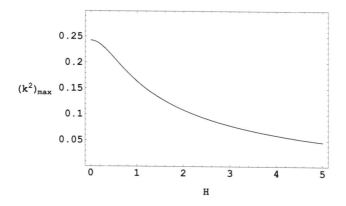

Fig. 7.3 Square k_{max}^2 of the wavenumber for maximum growth rate of the four-wave instability, as a function of H, calculated from a series expansion of the unstable root in (7.69) up to $O(k^6)$, according to [5]. Here $E_0 = 0.5$

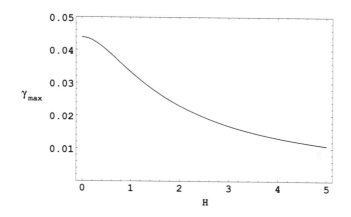

Fig. 7.4 Maximum growth rate γ_{max} as a function of the quantum parameter H according to [5]. Pump electric field amplitude: $E_0 = 0.5$

In the more general case when $k_0 \neq 0$, (7.64) needs to be numerically studied from the beginning. Figure 7.5 displays the real (solid lines) and imaginary (dashed lines) parts of ω as a function of k. Both uncoupled (i.e., $E_0 \approx 0$) and coupled cases are considered, for three different values of H. In view of the symmetry $(k, \omega) \leftrightarrow (-k, -\omega) \Rightarrow D_1 \leftrightarrow D_2$ of the dispersion relation (7.64), only positive wavenumbers are necessary. Of special relevance is the overlay region of the branches D_s and D_2, where instability occurs. In the uncoupled case, $k = 2k_0$ is a root of D_2 when $\omega = 0$, for both classical and quantum cases. In addition, the plots of D_s and D_2 branches touch each other at isolated points while, when $E_0 \neq 0$, overlay occurs in a finite interval $k \in I_k = (k_a, k_b)$ corresponding to instability. The first column of plots shows that, for a fixed k_0, both uncoupled curves raise with H, implying

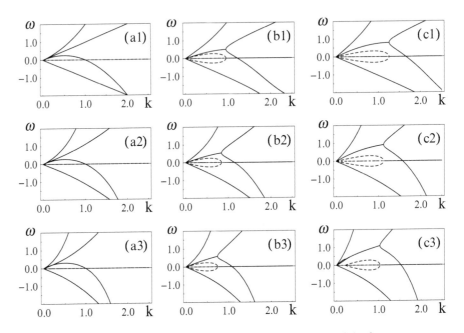

Fig. 7.5 Real (*solid lines*) and imaginary (*dashed lines*) components of the frequency ω as a function of k for uncoupled (frames a1–a3) and coupled cases (frames b1–c3) according to [5]. From *top* to *bottom*, $H = 0$, $H = 0.5$ and $H = 0.9$, respectively. From *left* to *right*, $E_0 = 0$, $E_0 = 0.5$ and $E_0 = 0.5$. For the first and second columns $k_0 = 0.5$; for the third column, $k_0 = 0.75$

reduction of the unstable interval I_k. The same tendency of a stretching unstable range of wavenumbers is apparent in the second and third columns.

For nonzero pump amplitude E_0, both the second and third columns of Fig. 7.5 shows an overall contraction of I_k. This results from the gradual shift of k_a to the right and k_b to the left, due to the quantum effects. Hence, the numerical results show that the quantum effect inhibits the efficient spreading of energy among different modes. Moreover, for a specific k, $N_I = (k_b - k_a)/k$ represents a measure of the number of active unstable modes. Hence, a shrinking I_k means that the Langmuir fluctuations in quantum plasmas represent more coherent configurations, that is, having less effective unstable modes in comparison with the classical limit.

7.3 Nonlinear Analysis

In the previous parametric instabilities analysis, only the linear content of the quantum Zakharov system was explored. To start investigating arbitrary amplitude waves, first notice that exact solutions for (7.38) and (7.39) are possible if we consider pure ion-sound waves, defined by $E = 0$. With zero electric field, the density perturbation satisfy

$$\frac{\partial^2 n}{\partial t^2} - \frac{\partial^2 n}{\partial x^2} + H^2 \frac{\partial^4 n}{\partial x^4} = 0, \tag{7.72}$$

which is the only remaining equation to be solved. In spite of the apparent simplicity, this linear fourth-order wave equation is not endowed with a rich symmetry structure. Indeed, searching for Lie point symmetries [15], the only invariance transformations we have found are time and space translation symmetries, as well as a scale symmetry associated with the linearity. In other words, the Lorentz invariance present in the case of classical ($H = 0$) ion sound waves is not available. A more detailed Lie symmetry analysis of the full ($E \neq 0$) quantum Zakharov equations was performed by Tang and Shukla [17]. In this work it was shown that the classical and quantum Zakharov systems admit the same symmetry groups. Similarity reductions were obtained and a pure general periodic ion-acoustic wave solution derived.

In the $E = 0$ case, exact traveling waves for (7.72) can be found supposing $n = \bar{n}(x - ct)$, for constant c and $\bar{n}$ a function to be determined. For supersonic flow ($c^2 > 1$), disregarding an integration constant associated with unbounded solutions we get periodic solutions given by

$$n = a + b \cos\left(\frac{\sqrt{c^2 - 1}}{H}(x - ct) + \delta\right), \tag{7.73}$$

where a, b, and δ are numerical constants. In the reference frame of the traveling wave, the wavenumber of the solution increase without bound as $H \to 0$. Hence, the classical limit of (7.73) is singular, as can be expected from the beginning since the order of (7.72) is changed from 4 to 2 as the quantum parameter goes to zero.

For nonzero electric field, some considerations on the fully nonlinear realm of (7.38) and (7.39) can be made at least in the adiabatic regime, where the density fluctuation is very slowly changing in time. In the classical adiabatic limit, the Zakharov system reduces to the nonlinear Schrödinger equation, which is a completely integrable model [1]. Hence, it is tempting to try to reproduce the classical procedure for the derivation of quasi-static solutions, setting $\partial^2 n / \partial t^2 \to 0$ in (7.39). Integration of this equation with, for example, decaying or periodic boundary conditions then gives

$$n = -|E|^2 + H^2 \frac{\partial^2 n}{\partial x^2}. \tag{7.74}$$

The strategy is not totally absurd, since the variables entering the quantum Zakharov system are slowly varying in time, so that the second-order time-derivative of n can with some confidence be taken as negligible. Inserting (7.74) into (7.38) yields

$$i\frac{\partial E}{\partial t} + \frac{\partial^2 E}{\partial x^2} + |E|^2 E = H^2\left(\frac{\partial^4 E}{\partial x^4} + E\frac{\partial^2 n}{\partial x^2}\right). \tag{7.75}$$

In the formal classical limit $H \to 0$, the right-hand side of (7.75) vanishes and we are left with the nonlinear Schrödinger equation, which is completely integrable by the inverse scattering transform. In this case, it has a number of properties including the existence of N-soliton solutions [1].

On the other hand, as argued in [5], it is interesting to consider the simultaneous semiclassical and adiabatic limits of the quantum Zakharov model. In this case, it is allowed to replace $n \to -|E|^2$ on the right-hand side of (7.75), which is already a quantum correction. We are then left with the decoupled equation

$$i\frac{\partial E}{\partial t} + \frac{\partial^2 E}{\partial x^2} + |E|^2 E = H^2\left(\frac{\partial^4 E}{\partial x^4} - E\frac{\partial^2 |E|^2}{\partial x^2}\right) \tag{7.76}$$

for the envelope electric field. Equation (7.76) can be used as a starting point to study first-order quantum perturbations of the classical soliton solutions. The generalized nonlinear Schrödinger equation (7.76) will be addressed in the next section in the light of an underlying Lagrangian formalism.

7.4 Semiclassical Adiabatic Regime

The nonlinear Schrödinger equation is completely integrable with N-soliton solutions, as a consequence of a detailed balance between dispersive and nonlinear contributions [1]. Therefore, we could ask at this point about the way the quantum effects perturb or perhaps even destroy these localized solitonic structures. Since quantum effects enhance dispersion in view of the Bohm potential, one should expect that solitons for the quantum Zakharov equations will be not easy to find. To investigate this conjecture consider first the simultaneous adiabatic and semiclassical case, which reduces to the decoupled (7.76) for E.

Equation (7.76) is derivable from a variational principle,

$$\delta S = \delta \int \mathscr{L}\,\mathrm{d}x\mathrm{d}t = 0, \tag{7.77}$$

with a Lagrangian density

$$\mathscr{L} = \frac{i}{2}\left(E^*\frac{\partial E}{\partial t} - E\frac{\partial E^*}{\partial t}\right) - \frac{\partial E^*}{\partial x}\frac{\partial E}{\partial x} + \frac{|E|^4}{2} - H^2\frac{\partial^2 E^*}{\partial x^2}\frac{\partial^2 E}{\partial x^2} + \frac{H^2}{2}|E|^2\frac{\partial^2 |E|^2}{\partial x^2}. \tag{7.78}$$

The variational derivatives $\delta S/\delta E^* = \delta S/\delta E = 0$ produce (7.76) and the complex conjugate equation, respectively.

We recall that for the higher-order Lagrangian density shown in (7.78) the variational derivatives are given by

$$\frac{\delta \mathscr{L}}{\delta E} = \frac{\partial \mathscr{L}}{\partial E} - \frac{\partial}{\partial t}\left(\frac{\partial \mathscr{L}}{\partial(\partial E/\partial t)}\right) - \frac{\partial}{\partial x}\left(\frac{\partial \mathscr{L}}{\partial(\partial E/\partial x)}\right) + \frac{\partial^2}{\partial x^2}\left(\frac{\partial \mathscr{L}}{\partial(\partial^2 E/\partial x^2)}\right),$$

(7.79)

$$\frac{\delta \mathscr{L}}{\delta E^*} = \frac{\partial \mathscr{L}}{\partial E^*} - \frac{\partial}{\partial t}\left(\frac{\partial \mathscr{L}}{\partial(\partial E^*/\partial t)}\right) - \frac{\partial}{\partial x}\left(\frac{\partial \mathscr{L}}{\partial(\partial E^*/\partial x)}\right) + \frac{\partial^2}{\partial x^2}\left(\frac{\partial \mathscr{L}}{\partial(\partial^2 E^*/\partial x^2)}\right).$$

(7.80)

The existence of a Lagrangian can be used to search for approximate solutions. In quantum mechanics a time-independent variational method consider searching for the minimal energy assuming a reasonable *Ansatz* for the eigenfunctions and calculating the expectation value of the Hamiltonian operator. The same methodology can be pursued for nonlinear time-dependent problems too, where trial wavefunctions are supposed to extremize an underlying action integral. This strategy yield better results if the choice of the trial wavefunction is dictated by physical and mathematical arguments. For instance, in the case of Bose–Einstein condensates described by a Gross–Pitaevskii equation with a parabolic confinement potential a time-dependent Gaussian *Ansatz* is the natural guess [9]. Indeed, the confinement certainly produce some localized solution. Moreover, neglecting the nonlinear interaction term the corresponding Gross–Pitaevskii equation becomes the (linear) Schrödinger equation for the simple harmonic oscillator, for which the ground state is a Gaussian.

In the strict classical case $H = 0$, (7.76) admit the one soliton solution

$$E = E_0 \exp\left(\frac{iE_0^2 t}{2}\right)\operatorname{sech}\left(\frac{E_0 x}{\sqrt{2}}\right),$$

(7.81)

where E_0 is a constant. Inspired by this result, we postulate the time-dependent, localized in space trial function

$$E = \alpha(t)\exp(i\theta(t))\operatorname{sech}(\beta(t)x),$$

(7.82)

for the quantum nonlinear Schrödinger equation (7.76), composed by adjustable real functions $\alpha(t), \beta(t)$, and $\theta(t)$, considered as functions of time only.

Inserting the *Ansatz* (7.82) into the functional (7.77) and performing the spatial integration, we get

$$S = \int L\,dt$$

(7.83)

with a Lagrange function

$$L = L(\theta, \alpha, \beta, \dot{\theta}, \dot{\alpha}, \dot{\beta}) = \frac{\alpha^2}{15} \left(\frac{30\dot{\theta}}{\beta} - \frac{10\alpha^2}{\beta} + 10\beta + 8H^2\alpha^2\beta + 14H^2\beta^3 \right).$$
(7.84)

With the spatial dependence of the solution defined from advance, there remains a finite-dimensional mechanical system where all quantities are functions of time only. Since the action should be extremized, the variational derivatives of the Lagrangian (7.84) equals zero, taking α, β and θ as independent variables. Therefore,

$$\frac{\delta S}{\delta \theta} = \frac{\partial L}{\partial \theta} - \frac{d}{dt} \left(\frac{\partial L}{\partial \dot{\theta}} \right) = 0,$$
(7.85)

$$\frac{\delta S}{\delta \alpha} = \frac{\partial L}{\partial \alpha} - \frac{d}{dt} \left(\frac{\partial L}{\partial \dot{\alpha}} \right) = 0,$$
(7.86)

$$\frac{\delta S}{\delta \beta} = \frac{\partial L}{\partial \beta} - \frac{d}{dt} \left(\frac{\partial L}{\partial \dot{\beta}} \right) = 0.$$
(7.87)

Computing the variational derivatives, we obtain

$$\frac{\delta S}{\delta \theta} = 0 \Rightarrow \frac{d}{dt} \left(\frac{\alpha^2}{\beta} \right) = 0,$$
(7.88)

$$\frac{\delta S}{\delta \alpha} = 0 \Rightarrow \alpha \left(\dot{\theta} - \frac{2\alpha^2}{3} + \frac{\beta^2}{3} + \frac{8H^2\alpha^2\beta^2}{15} + \frac{7H^2\beta^4}{15} \right) = 0,$$
(7.89)

$$\frac{\delta S}{\delta \beta} = 0 \Rightarrow \alpha^2 \left(\dot{\theta} - \frac{\alpha^2}{3} - \frac{\beta^2}{3} - \frac{4H^2\alpha^2\beta^2}{15} - \frac{7H^2\beta^4}{5} \right) = 0.$$
(7.90)

Equation (7.88) imply

$$\alpha^2 = \sqrt{2}E_0\beta,$$
(7.91)

where E_0 is a numerical constant. Excluding the trivial case $\alpha = 0$ and inserting (7.91) into (7.89), we derive

$$\dot{\theta} = \frac{2\sqrt{2}E_0\beta}{3} - \frac{\beta^2}{3} - \frac{8\sqrt{2}E_0H^2\beta^3}{15} - \frac{7H^2\beta^4}{15},$$
(7.92)

which, taking into account (7.90), yields

$$H^2\beta^3 + \frac{3\sqrt{2}E_0H^2\beta^2}{7} + \frac{5\beta}{14} - \frac{5\sqrt{2}E_0}{28} = 0.$$
(7.93)

Equations (7.91)–(7.93) show that α and β are constants while θ is a linear function of t, just like in the classical case. However, it is worth to consider the dependence of the several expressions on H, to see what is the rôle of the quantum effects on the one soliton solution (7.81). For this purpose (7.93) is essential. An useful simplification of it is achieved through the rescaling

$$\bar{\beta} = \frac{\sqrt{2}\beta}{E_0}, \quad \bar{H} = \frac{\sqrt{2}E_0 H}{2}, \tag{7.94}$$

which eliminates one free parameter,

$$\bar{H}^2\bar{\beta}^3 + \frac{6\bar{H}^2\bar{\beta}^2}{7} + \frac{5\bar{\beta}}{14} - \frac{5}{14} = 0. \tag{7.95}$$

In the formal classical limit $\bar{H} = 0$, (7.95) imply $\bar{\beta} = 1$, recovering the classical one soliton solution. Moreover, considering the discriminant of (7.95) it follows that for

$$\bar{H}^2 \leq \frac{5}{1,152}(681 + 23\sqrt{897}) \approx 5.946, \tag{7.96}$$

it admits only one real besides two complex conjugate roots. This range of parameters is in line with the semiclassical limit. However, it is important to notice that, in view of the dependence of $\bar{H}$ on E_0 (see (7.94)), no constraint is imposed on the maximum value of $\bar{H}$, since sufficiently high values of E_0 can be freely chosen. Hence, we now consider both large and small values of $\bar{H}$.

7.4.1 Small $\bar{H}^2$

For $\bar{H} \ll 1$, solving (7.95) recursively gives

$$\bar{\beta} = 1 - \frac{26\bar{H}^2}{5} + O(\bar{H}^4). \tag{7.97}$$

Retaining only the leading quantum correction and transforming back to the original variables yield

$$\alpha = E_0 \left(1 - \frac{13\Omega H^2}{5}\right), \tag{7.98}$$

$$\beta = \frac{\sqrt{2}E_0}{2}\left(1 - \frac{26\Omega H^2}{5}\right), \tag{7.99}$$

$$\dot{\theta} = \frac{E_0^2}{2} - 5\frac{E_0^4}{4}H^2. \tag{7.100}$$

By inspection of the variational solution (7.82), we conclude that the amplitude and the rate of change the phase θ became smaller due to quantum effects, while the spatial extent of the soliton has the opposite behavior. Hence, the quantum diffraction effects tend to enlarge the dispersion of the soliton, destroying localization. This is similar to the well-known quantum mechanical effect of wave packet spreading, a signature of the indeterminacy principle.

7.4.2 Large $\bar{H}^2$

Now consider the case $\bar{H} = \sqrt{2}E_0 H/2 \gg 1$, which can be accessed even for $H \ll 1$ provided H is not strictly zero and E_0 is sufficiently large. The leading order dependence of the terms in (7.95) can be examined supposing a power law expression $\beta \sim \bar{H}^r$ for some constant r. In this way, two possible families are found: (a) for $r = 0$ the first and second terms in (7.95) are the leading order contributions and can balance each other; (b) for $r = -1$ the first and third terms are the most relevant.

Proceeding in this way we find, instead of only one class of solutions as in the small $\bar{H}$ situation, three different subclasses, corresponding to

$$\bar{\beta}_0 = -\frac{6}{7} + \frac{65}{72}\bar{H}^{-2} + O\left(\bar{H}^{-4}\right), \tag{7.101}$$

$$\bar{\beta}_+ = \left(\frac{5}{12}\right)^{1/2}\bar{H}^{-1} - \frac{65}{144}\bar{H}^{-2} + O\left(\bar{H}^{-3}\right), \tag{7.102}$$

$$\bar{\beta}_- = -\left(\frac{5}{12}\right)^{1/2}\bar{H}^{-1} - \frac{65}{144}\bar{H}^{-2} + O\left(\bar{H}^{-3}\right). \tag{7.103}$$

The first solution comes from an expansion in $1/\bar{H}^2$, while the last two result from an expansion in $1/\bar{H}$.

Inspection of (7.91) shows a purely imaginary α is obtained from the leading terms in (7.101) or (7.103). However, this contradict the proposal (7.82), where α, β and θ are supposed real. Hence, we discard (7.101) and (7.103) and consider only $\bar{\beta}_+$ in (7.102). This yields, again considering only the leading order terms,

$$\alpha = \left(\frac{5E_0^2}{6H^2}\right)^{1/4}, \quad \beta = \frac{(5/12)^{1/2}}{H}, \quad \dot{\theta} = \frac{4}{9}\sqrt{\frac{5}{6}}\frac{E_0}{H}. \tag{7.104}$$

Since $H \ll 1$ and $E_0 \gg 1$, this corresponds to a large amplitude, highly localized, and highly oscillating variational solution, with no classical correspondence. Such approximate purely quantum periodic structure resembles the ion-sound nonlinear solution (7.73), which also has a singular limit when $H \to 0$.

In conclusion, quantum effects were shown to contribute to destroy localized structures of the adiabatic semiclassical limit. The employed time-dependent variational formalism also points to a new, strongly oscillating pattern as described by (7.82) and (7.104). Many open questions still remain, in particular on the rôle of quantum effects on the stability of the N-soliton solutions admitted in the classical limit. These N-soliton solutions were found, through numerical experiments [7], to typically act as asymptotic equilibria in the classical case. Even if in principle one can expect that quantum terms would tend to eliminate any localized structure due to (roughly speaking) wave packet diffraction, this is not an obvious property since in quantum plasmas the nonlinearities may produce unexpected coherent phenomena. For instance, nonlinear structures can appear in quantum electron plasmas, in the form of dark solitons and vortices. These structures were numerically shown to stable, and therefore useful for the transport of information at quantum scales [4].

The time-dependent variational method is an useful method to explore the nonlinear behavior of systems of evolution equations, provided a Lagrangian formalism is available. In the next section, we formulate the one-dimensional quantum Zakharov system as a Lagrangian system, regardless of adopting the adiabatic and semiclassical limit.

7.5 Time-Dependent Variational Method

Coherent structures like solitons, vortices, cavitons, spikons and so on play a relevant rôle in any physical system. In quantum plasmas, the numerical stability of vortices and dark solitons has been detected in Schrödinger–Poisson quantum electron plasmas [4]. At the quantum scales, the transport of information in ultracold micromechanical systems can be addressed through such nonlinear structures. In this section, the influence of quantum effects on localized solutions of the quantum Zakharov system is considered in terms of an associated variational formalism and a trial function method. No restrictions to an adiabatic or semiclassical limit is imposed, as was done in the last section. Approximate methods are justified, since only few exact solutions are available for the quantum Zakharov system [17].

In the classical case, the Zakharov equations admit a Langmuir soliton solution [18]. The internal vibrations of such solitary waves were analyzed by a variational approach using Gaussian trial functions [12], thanks to the associated Lagrangian formalism. Time-dependent variational methods are also traditional, for instance, in the study of nonlinear pulse propagation in optical fibers [2] and Bose–Einstein condensates [9]. Hence, it is a natural trend to search for a Lagrangian description and then using it in the stability analysis of localized waves for the quantum Zakharov system. In other words, using a Gaussian *Ansatz* or similar solitary pulses as a trial function extremizing an action functional, we can derive information about the perturbation of the Langmuir soliton by quantum effects. We can expect the enlargement of the width and a smaller amplitude of these pulses, due to wave packet spreading. This tendency has been verified in the adiabatic and semiclassical case in the preceding section. However, it can also happens that pure quantum instabilities tend to destroy the localized structures, beyond a simple deformation. Moreover, in a mathematical ground the construction of a variational formulation for the quantum Zakharov equations is important in itself.

We can be more specific. Indeed the quantum Zakharov contains a dimensionless parameter H representative of the strength of the quantum effects, see (7.37). Treating H as a control parameter, a variational method can be used to access the modifications of a trial Gaussian profile as H varies. In this way, a general tendency can be understood, even without the knowledge of any exact solution.

The one-dimensional quantum Zakharov equations (7.38) and (7.39) are derivable from the Lagrangian density [11]

$$\mathscr{L} = \frac{i}{2}\left(E^*\frac{\partial E}{\partial t} - E\frac{\partial E^*}{\partial t}\right) - \left|\frac{\partial E}{\partial x}\right|^2 - \frac{\partial u}{\partial x}|E|^2 + \frac{1}{2}\left(\frac{\partial u}{\partial t}\right)^2 - \frac{1}{2}\left(\frac{\partial u}{\partial x}\right)^2$$

$$-H^2\left|\frac{\partial^2 E}{\partial x^2}\right|^2 - \frac{H^2}{2}\left(\frac{\partial^2 u}{\partial x^2}\right)^2, \tag{7.105}$$

where it was introduced the auxiliary variable u from which the density can be found,

$$n = \frac{\partial u}{\partial x}. \tag{7.106}$$

The Lagrangian densities in (7.78) and (7.105) can be compared. In the general, nonadiabatic and nonsemiclassical case, one more variable need to be incorporated, since now the density is an independent quantity. Equation (7.105) is reminiscent from the classical Lagrangian density [6], with the appropriate quantum contributions added.

The independent fields are E, E^*, and u. Besides (7.4) and (7.4) one more variational derivative should be calculated. In the present case, it is

$$\frac{\delta\mathscr{L}}{\delta u} = -\frac{\partial}{\partial t}\left(\frac{\partial\mathscr{L}}{\partial(\partial u/\partial t)}\right) - \frac{\partial}{\partial x}\left(\frac{\partial\mathscr{L}}{\partial(\partial u/\partial x)}\right) + \frac{\partial^2}{\partial x^2}\left(\frac{\partial\mathscr{L}}{\partial(\partial^2 u/\partial x^2)}\right). \tag{7.107}$$

Inserting $\mathscr{L}$ from (7.105) into (7.4), (7.4), and (7.107), we get

$$\frac{\delta\mathscr{L}}{\delta E} = 0 \Rightarrow -i\frac{\partial E^*}{\partial t} + \frac{\partial^2 E^*}{\partial x^2} - H^2\frac{\partial^4 E^*}{\partial x^4} = \frac{\partial u}{\partial x}E^*, \tag{7.108}$$

$$\frac{\delta\mathscr{L}}{\delta E^*} = 0 \Rightarrow i\frac{\partial E}{\partial t} + \frac{\partial^2 E}{\partial x^2} - H^2\frac{\partial^4 E}{\partial x^4} = \frac{\partial u}{\partial x}E, \tag{7.109}$$

$$\frac{\delta\mathscr{L}}{\delta u} = 0 \Rightarrow -\frac{\partial}{\partial x}\left(|E|^2 + \frac{\partial u}{\partial x}\right) + \frac{\partial^2 u}{\partial t^2} + H^2\frac{\partial^4 u}{\partial x^4} = 0. \tag{7.110}$$

After differentiation with respect to x, the last equation reproduces (7.39).

The classical Zakharov system is nonintegrable, except in the adiabatic limit when it reduces to the nonlinear Schrödinger equation which is tractable by the inverse scattering transform. However, it admits [18] some analytical solutions, among which we select the Langmuir soliton

$$E = E_0 \exp \left(\frac{i E_0^2 t}{2} \right) \operatorname{sech} \left(\frac{E_0 x}{\sqrt{2}} \right),   \tag{7.111}$$

$$n = -E_0^2 \operatorname{sech}^2 \left(\frac{E_0 x}{\sqrt{2}} \right),   \tag{7.112}$$

where E_0 is an arbitrary real parameter. Notice that (7.111) provides a solution for the corresponding nonlinear Schrödinger equation too, see (7.81). Strictly, the expression "Langmuir soliton" is not applicable since different solutions (7.111) and (7.112), for different parameter E_0, do not produce simple phase shifts under collisions [18], as expected for solitons. Nevertheless, we follow the traditional usage in the spirit that the coherent waves in (7.111) and (7.112) results from the equilibrium between nonlinearity and dispersion, in the same way as solitons form. Physically, the Langmuir soliton represents a hole in the low frequency part of the electron–ion density maintained self-consistently by the ponderomotive force.

Isolated classical Langmuir solitons are remarkably stable and do not decay [18]. What is the effect of quantum perturbations on such structures? To answer the question on analytic grounds we can use the variational formalism and set a time-dependent trial function reproducing the gross features of (7.111) and (7.112). In addition, it is desirable to have a sufficiently tractable *Ansatz*. These requirements are satisfied by the Gaussian profile

$$E = A \exp \left(-\frac{x^2}{2a^2} + i\phi + i\kappa x^2 \right),   \tag{7.113}$$

$$n = -B \exp \left(-\frac{x^2}{b^2} \right),   \tag{7.114}$$

where A, B, a, b, ϕ, and κ are functions of time only. We assume A and B positive to maintain resemblance with (7.111) and (7.112). The "chirp function" κ is responsible for the increase of the spatial frequency of oscillations of the envelope electric field for larger distances, since $E \sim \exp(i\kappa x^2)$. It is not present in the original Langmuir soliton, but is a necessary mathematical ingredient in the Lagrangian formalism.

As for any trial function, the main drawback of the Gaussian profile is that it does not modify its spatial shape as time evolves. The advantage is the transformation of a set of nonlinear partial differential equations in a set of nonlinear ordinary differential equations, whose properties can be more easily accessed. The classical Zakharov system can also be treated by a variational approach using a combination of Jacobi elliptic functions [16], but we use Gaussian functions for the sake of simplicity. In the same spirit (7.113) and (7.114) are analytically simpler than other localized forms involving hyperbolic functions, Lorentzian distributions and so on.

To calculate the Lagrangian

$$L = \int_{-\infty}^{\infty} \mathscr{L} \, dx   \tag{7.115}$$

corresponding to the profile (7.113) and (7.114) there is the need of the auxiliary quantity $u(x,t)$. From (7.106) and (7.114), we have

$$u = u_0(t) - Bb\frac{\sqrt{\pi}}{2}\mathrm{Erf}\left(\frac{x}{b}\right), \tag{7.116}$$

involving an arbitrary function $u_0(t)$ depending on time only and the error function defined by

$$\mathrm{Erf}(x) = \frac{2}{\sqrt{\pi}} \int_0^x \exp(-x'^2)\mathrm{d}x'. \tag{7.117}$$

It turns out that taking into account (7.116) and computing $\partial u/\partial x$ and $\partial u/\partial t$ the Lagrangian (7.115) diverges except for constant u_0 and $Bb = M$, where M is a constant. Without loss of generality we can take $u_0 = 0$.

The invariance of $M = Bb$ is in line with the conservation of the low frequency part of the "mass"

$$M \equiv -\frac{1}{\sqrt{\pi}} \int_{-\infty}^{+\infty} n\mathrm{d}x = Bb, \tag{7.118}$$

the last equality following from the variational solution and with the factor $1/\sqrt{\pi}$ being introduced for convenience. The minus sign assures a positive value. In other words, imposing $B = M/b$ is equivalent to the invariance of the low frequency part of the mass. Actually M could be better associated with an absence of mass, since we are considering hole, or caviton solutions.

In addition, the quantum Zakharov equations admit the conservation laws for the number of high frequency quanta

$$N = \frac{1}{\sqrt{\pi}} \int_{-\infty}^{+\infty} |E|^2\mathrm{d}x, \tag{7.119}$$

for the linear momentum

$$P = \frac{i}{2} \int_{-\infty}^{+\infty} \left(E^*\frac{\partial E}{\partial x} - \frac{\partial E^*}{\partial x}E\right) \mathrm{d}x, \tag{7.120}$$

and for the energy

$$\mathscr{H} = \frac{1}{\sqrt{\pi}} \int_{-\infty}^{\infty} \left[\left|\frac{\partial E}{\partial x}\right|^2 + \frac{\partial u}{\partial x}|E|^2 + \frac{1}{2}\left(\frac{\partial u}{\partial t}\right)^2 + \frac{1}{2}\left(\frac{\partial u}{\partial x}\right)^2\right] \mathrm{d}x. \tag{7.121}$$

In other words,

$$\dot{N} = 0, \quad \dot{P} = 0, \quad \dot{\mathscr{H}} = 0. \tag{7.122}$$

The conservation laws can be directly checked using the quantum Zakharov equations.

The conserved quantities for the quantum Zakharov system can be found sistematically from the application of Noether's theorem, relating symmetries and conservation laws [15]. In this context, the invariance of N, P and $\mathcal{H}$ are associated with the symmetry of the action integral under a phase transformation of the electric field and space and time translations, respectively.

For instance, the energy integral exist since the Lagrangian density (7.105) is invariant under time translation, with the conserved quantity being the Hamiltonian. To compute the Hamiltonian we consider the momenta Π_E, Π_{E^*}, and Π_u associated with E, E^*, and u,

$$\Pi_E = \frac{\partial \mathcal{L}}{\partial(\partial E/\partial t)} = \frac{i}{2}E^*, \tag{7.123}$$

$$\Pi_{E^*} = \frac{\partial \mathcal{L}}{\partial(\partial E^*/\partial t)} = -\frac{i}{2}E, \tag{7.124}$$

$$\Pi_u = \frac{\partial \mathcal{L}}{\partial(\partial u/\partial t)} = \frac{\partial u}{\partial t} \tag{7.125}$$

and take the Legendre transform as usual,

$$\mathcal{H} = \frac{1}{\sqrt{\pi}} \int_{-\infty}^{\infty} \left(\Pi_E \frac{\partial E}{\partial t} + \Pi_{E^*} \frac{\partial E^*}{\partial t} + \Pi_u \frac{\partial u}{\partial t} - \mathcal{L} \right) dx. \tag{7.126}$$

The factor $1/\sqrt{\pi}$ is just a matter of convenience.

In addition, for the (classical) Langmuir soliton in (7.111) and (7.112), we have the property

$$|E|^2 = -n. \tag{7.127}$$

To assure that the Gaussian form in (7.113) and (7.114) reproduces somehow the Langmuir soliton, we postulate that

$$\frac{1}{\sqrt{\pi}} \int_{-\infty}^{\infty} |E|^2 dx = -\frac{1}{\sqrt{\pi}} \int_{-\infty}^{\infty} n \, dx, \tag{7.128}$$

so as to satisfy (7.127) in a global sense. Equivalently, from (7.118) and (7.119) we impose

$$M = N, \tag{7.129}$$

in what follows, eliminating one free parameter. The above equation express a sort of balance: the number of quanta originates from the density depletion. In the remaining, we examine only the balanced solutions for which N equals the low frequency part of the mass.

For the proposed trial functions,

$$N = A^2 a, \tag{7.130}$$

$$P = 0, \tag{7.131}$$

$$\mathcal{H} = \frac{M}{2}\left(\frac{1}{a^2} + 4\kappa^2 a^2\right) - \frac{M^2}{\sqrt{a^2+b^2}} + \frac{M^2}{2\sqrt{2}b} + \frac{M^2 b^2}{8\sqrt{2}b}$$

$$+ \frac{3H^2 M}{4}\left(\frac{1}{a^2} + 4\kappa^2 a^2\right)^2 + \frac{H^2 M^2}{2\sqrt{2}b^3}. \tag{7.132}$$

The invariance of N and H imply useful simplifications, while the momentum conservation is not significant here since for the present *Ansatz* the "center-of-mass" of the Langmuir soliton is nontranslating, or $P = 0$.

Observe that for the adiabatic semiclassical case the generalized nonlinear Schrödinger equation (7.76) admit the same N and P invariants as shown in (7.119) and (7.120), while the corresponding energy integral needs to be adapted to

$$\mathcal{H} = \frac{1}{\sqrt{\pi}}\int_{-\infty}^{\infty}\left[\left|\frac{\partial E}{\partial x}\right|^2 - \frac{1}{2}|E|^2 + H^2\left|\frac{\partial^2 E}{\partial x^2}\right|^2 - \frac{H^2|E|^2}{2}\frac{\partial^2}{\partial x^2}\left(|E|^2\right)\right]dx, \tag{7.133}$$

not containing the auxiliary variable u.

All the ingredients for the evaluation of the Lagrangian L in (7.115) are known. Using (7.105), (7.113), (7.114) and (7.116), after spatial integration, we get

$$L = \sqrt{\pi}\left[-A^2 a\dot{\phi} - \frac{A^2 a}{2}\left(a^2\dot{\kappa} + \frac{1}{a^2} + 4\kappa^2 a^2\right) + \frac{MA^2 a}{\sqrt{a^2+b^2}} - \frac{M^2}{2\sqrt{2}b}\right.$$

$$\left.+ \frac{M^2 b^2}{8\sqrt{2}b} - \frac{3H^2 A^2 a}{4}\left(\frac{1}{a^2} + 4\kappa^2 a^2\right)^2 - \frac{H^2 M^2}{2\sqrt{2}b^3}\right], \tag{7.134}$$

depending on the dynamical variables ϕ, κ, A, a, b and their derivatives. Hence, we have a five-dimensional configuration space. Only the last two terms in (7.134) contain quantum corrections.

The Euler–Lagrange equations follow in the usual way, since L is depending only on the dynamical variables and their first-order time-derivatives. For instance, we have

$$\frac{\partial L}{\partial \phi} - \frac{d}{dt}\left(\frac{\partial L}{\partial \dot{\phi}}\right) = 0 \Rightarrow A^2 a = N = \text{constant}, \tag{7.135}$$

just reproducing the conservation of the number N of high frequency quanta, see (7.130). Similarly, virtual variations of κ gives

$$\frac{\partial L}{\partial \kappa} - \frac{d}{dt}\left(\frac{\partial L}{\partial \dot{\kappa}}\right) = 0 \Rightarrow a\dot{a} = 4\kappa a^2 + 12H^2\kappa(1 + 4\kappa^2 a^4). \tag{7.136}$$

In contrast to the classical case where the chirp function κ is easily derived from $a(t)$, in the quantum case κ is the solution of the third-degree equation (7.136). Expressing κ in terms of a and $\dot{a}$ would give too cumbersome equations. A reasonable alternative in the semiclassical limit is to solve (7.136) for κ as a power series in H^2. Other possibility is to regard (7.136) as a dynamical equation to be included in numerical simulations. Let us consider both approaches.

The system of Euler–Lagrange equations for a and A is equivalent to a linear system for $\dot{\kappa}$ and $\dot{\phi}$, because L does not depend on the derivatives of a and A:

$$\frac{\partial L}{\partial A} - \frac{d}{dt}\left(\frac{\partial L}{\partial \dot{A}}\right) = 0 \Rightarrow \dot{\phi} + \frac{1}{2}\left(a^2\dot{\kappa} + \frac{1}{a^2} + 4\kappa^2a^2\right)$$

$$-\frac{M}{\sqrt{a^2+b^2}} + \frac{3H^2}{4}\left(\frac{1}{a^2} + 4\kappa^2a^2\right)^2 = 0, \qquad (7.137)$$

$$\frac{\partial L}{\partial a} - \frac{d}{dt}\left(\frac{\partial L}{\partial \dot{a}}\right) = 0 \Rightarrow \dot{\phi} + \frac{1}{2}\left(3a^2\dot{\kappa} - \frac{1}{a^2} + 12\kappa^2a^2\right)$$

$$-\frac{Mb^2}{(a^2+b^2)^{3/2}} + 3H^2\left(-\frac{3}{4a^4} + 2\kappa^2 + 20\kappa^4a^4\right) = 0. \quad (7.138)$$

Eliminating $\dot{\phi}$, we get

$$a\dot{\kappa} = \frac{1}{a^3} - 4\kappa^2a - \frac{Ma}{(a^2+s^4)^{3/2}} + \frac{3H^2}{a^5} - 48H^2\kappa^4a^3. \qquad (7.139)$$

Finally, variation of b gives

$$\frac{\partial L}{\partial b} - \frac{d}{dt}\left(\frac{\partial L}{\partial \dot{b}}\right) = 0 \Rightarrow \ddot{s} = \frac{1}{s^3} - \frac{2\sqrt{2}s^3}{(a^2+s^4)^{3/2}} + \frac{3H^2}{s^7}, \qquad (7.140)$$

in terms of a new variable $s = \sqrt{b}$ and using the conservation law in (7.130) to eliminate A.

Equations (7.136), (7.139) and (7.140) form a closed system for the dynamical variables $a, \kappa,$ and s and are the basis for what follows. The small and large values of H can be analyzed separately.

7.5.1 The Small H Case

When $H \ll 1$ is a small parameter, (7.136) can be solved to first-order in H^2 yielding

$$\kappa = \frac{\dot{a}}{4a} - H^2\left(\frac{3\dot{a}}{4a^3} + \frac{3\dot{a}^3}{16a}\right), \qquad (7.141)$$

disregarding higher-order quantum corrections. Inserting κ in (7.139) one is left with a coupled, nonlinear system of second-order equations for a and s, namely, (7.140) and

$$\ddot{a} = \frac{4}{a^3} - \frac{4Ma}{(a^2 + s^4)^{3/2}} + H^2 \left(\frac{24}{a^5} + \frac{6\dot{a}^2}{a^3} - \frac{12M}{a(a^2 + s^4)^{3/2}} - \frac{9Ma\dot{a}^2}{(a^2 + s^4)^{3/2}} \right). \quad (7.142)$$

The energy (7.132), evaluated using (7.141), is

$$\mathcal{H} = \frac{1}{4} \left(M \left[\frac{\dot{a}^2}{2} + \frac{2}{a^2} \right] + \sqrt{2} M^2 \left[\dot{s}^2 + \frac{1}{s^2} \right] - \frac{4M^2}{\sqrt{a^2 + s^4}} \right)$$
$$+ \frac{H^2}{64} \left(\frac{48M}{a^4} - \frac{24M\dot{a}^2}{a^2} - 9M\dot{a}^4 + \frac{16\sqrt{2}M^2}{s^6} \right). \quad (7.143)$$

It is constant in the context of the semiclassical approximation:

$$\frac{\mathrm{d}\mathcal{H}}{\mathrm{d}t} = O(H^4) \quad (7.144)$$

along trajectories of (7.140) and (7.142). The invariance of $\mathcal{H}$ is a useful tool to check the accuracy of numerical schemes.

Equations (7.140) and (7.142) describe the nonlinear oscillations of the Gaussian widths of the electric field and the density depletion. Since we have a two degrees-of-freedom system, some simple tools from differential equations theory can be used to infer the behavior of the solutions, according to the parameters H and M. For instance, consider the linear stability analysis of the fixed points admitted by (7.140) and (7.142).

We can search for fixed points for the dynamical system as a power series in H^2. Setting the first- and second-order time-derivatives to zero in (7.140) and (7.142), we find the critical points at $(a, s) = (a_0, s_0)$, where

$$a_0 = \frac{2\sqrt{2}}{M} + \frac{3H^2 M}{\sqrt{2}}, \quad (7.145)$$

$$s_0 = \left(\frac{2\sqrt{2}}{M} \right)^{1/2} + \frac{H^2 M^{3/2}}{2^{1/4}}, \quad (7.146)$$

disregarding $O(H^4)$ terms. In terms of the original variables a and b these fixed points corresponds to Gaussians of same width at the formal classical limit. Quantum corrections, however, introduce a disturbance: the width a_0 of the equilibrium envelope electric field and the width $b_0 \sim s_0^2$ associated with the density start behaving differently. However, both characteristic lengths a_0 and b_0 increase with H, pointing for a wave packet spreading effect.

Consider small deviations $\sim \exp(i\omega t)$ from the equilibrium point, setting

$$a = a_0 + a_1 \exp(i\omega t), \quad s = s_0 + s_1 \exp(i\omega t) \tag{7.147}$$

where a_1 and b_1 are first-order quantities. Linearizing, one obtain

$$\omega^2 = \frac{M^2}{128}\left[24 + 10M^2 - H^2M^2(18M^2 + 33)\right]$$

$$\pm \frac{M^2}{64}\left[144 + 24M^2 + 25M^4 - H^2M^2(396 + 177M^2 + 90M^4)\right]^{1/2}. \tag{7.148}$$

For consistency one could also expand the square root at (7.148) up to $O(H^2)$. Then four possible frequency modes are detected, corresponding to the two possible choices of sign in (7.148). Instability comes when $\omega^2 < 0$. A detailed calculation in this way shows that instability will arrive when

$$H^2 > f(M^2) \equiv \frac{(24 + 10M^2)\sqrt{144 + 24M^2 + 25M^4} - 2(144 + 24M^2 + 25M^4)}{M^2((18M^2 + 33)\sqrt{144 + 24M^2 + 25M^4} - (396 + 177M^2 + 90M^4))}. \tag{7.149}$$

Notice that the internal soliton oscillations are stable in the classical case $H = 0$ for which always $\omega^2 > 0$. The expression in (7.149) differs from the result in [11] because here all quantities are expanded in powers of H. Nevertheless, the results are basically the same at the end.

The internal soliton vibrations can be unstable for small quantum parameter. For instance, for $M > 0.81$, $f(M^2) < 1$ at (7.149). Further increasing M allows for smaller values of H. For $M = 1$, one has $H > 0.79$ for instability, which is still reasonably attained at the underlying semiclassical limit. Too large values of M are not reasonable, however, due to the weak turbulence condition which precludes very strong electric fields. Figure 7.6 shows the marginal curve $H^2 = f(M^2)$ separating stable and unstable regions.

The presence of instabilities when $H \neq 0$ shows that the classical localized solution eventually disappear due to quantum effects, since the width of the corresponding Gaussian continuously increases with time. Hence, the classical Langmuir soliton, produced by particle trapping in the self-consistent electrostatic potential, becomes unstable due to electron tunneling. Differently from the usual tunneling process, here the electrons tunnel through a self-consistent potential barrier. At the end both the electron–hole and the electric field diffuse away. It can be the case, however, that nonlinear effects not included in the linear stability analysis above prevent the complete erasure of the Langmuir soliton.

Differently from the classical internal Langmuir soliton oscillations [12], (7.140) and (7.142) seems to be not described by a pseudo-potential function. Indeed due to the explicit velocity dependence on the dynamical equation for $a(t)$, we do not have an obvious recipe to express this system in a variational form. Nevertheless, using the energy integral $\mathcal{H}$ in (7.143) some conclusions can be obtained. For instance,

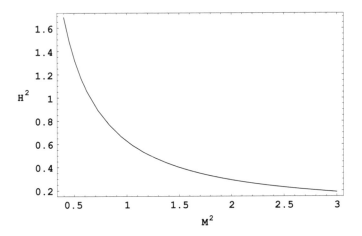

Fig. 7.6 The the marginal stability condition $H^2 = f(M^2)$ from (7.149). Instability happens for $H^2 > f(M^2)$

one step ahead the above linear analysis comprises a rough estimate for the escape velocity, defined as the minimum speed for unbounded motion. Asymptotically far from the self-consistent potential well, one has $\dot{a} = \dot{s} = 0$, in the unbounded case when $a \to \infty$ and $s \to \infty$. From (7.143) this correspond to $\mathcal{H} = 0$. Supposing an initial condition at the fixed point in (7.145) and (7.146) and defining $\dot{a}(0) = \dot{a}_0 = 0$ for the sake of simplicity, one find $\mathcal{H} = 0$ for a escape velocity $\dot{s}(0) = \dot{s}_0$ specified by

$$\dot{s}_0^2 = \frac{M}{4\sqrt{2}} - \frac{7H^2M^3}{64\sqrt{2}}, \qquad (7.150)$$

where $M = N$ was used. Quantum effects act in a tunneling-like manner once again because (7.150) shows that for $H > (4/\sqrt{7})M^{-1}$ the pseudo-particle with coordinates (a,s) will escape from (a_0,s_0) with $\dot{a}_0 = 0$ whatever the value of $\dot{s}_0$. The limiting value can be achieved even for the semiclassical case for sufficiently high M. For instance, when $M = 2$ the particle certainly escapes for $H > 0.76$, which can be regarded as moderate with some optimism. However, smaller values of H can be found for nonzero $\dot{a}_0$. In all cases, (7.150) confirm the overall diffusive effect of the Bohm potential in quantum plasmas.

Figures 7.7–7.9 shows typical oscillations for (7.140) and (7.142), describing the dynamics of the widths of the Gaussians associated with the electric field and the density, with $M = 3$, $H = 0.3$. The initial condition is at the fixed point and $\dot{a}(0) = 0$. Also, $\dot{s}_0 = 0.62$. For such parameters, simulations shows unbounded motion for $\dot{s}_0 = 0.64$, which is much less than the classical escape velocity, $\dot{s}_0 = 0.73$, and in good agreement with the critical value 0.59 arising from the crude estimate in (7.150). The energy $\mathcal{H}$ in (7.143) remains approximately constant at the value -0.10 along the run. We observe distinct time scales for $a(t)$ and $s(t)$. Taking a smaller initial value $\dot{s}_0$ gives a more regular, quasi-periodic oscillation

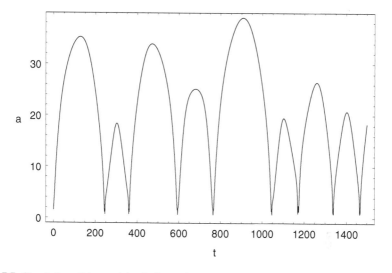

Fig. 7.7 Simulation of the semiclassical equations (7.140) and (7.142) showing $a(t)$, according to [11]. Parameters, $M = 3$, $H = 0.3$. Initial condition, $(a_0, s_0, \dot{a}_0, \dot{s}_0) = (1.52, 1.36, 0, 0.62)$

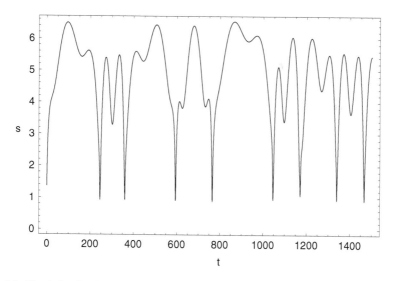

Fig. 7.8 Simulation for the semiclassical equations (7.140) and (7.142) showing $s(t)$, according to [11]. Parameters, $M = 3$, $H = 0.3$. Initial condition, $(a_0, s_0, \dot{a}_0, \dot{s}_0) = (1.52, 1.36, 0, 0.62)$

pattern, similar to the classical oscillations [12]. However, quantum effects imply complicated trajectories, approaching the critical value of $\dot{s}_0$ for unbounded motion (see Fig. 7.9). Additional runs show that increasing the value of $\dot{s}_0$ increases the period and the amplitude of the oscillations.

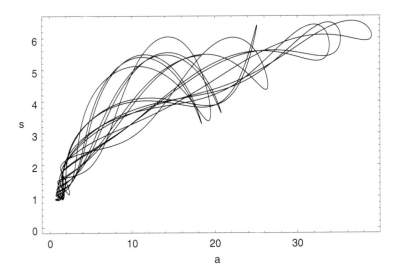

Fig. 7.9 Trajectory from (7.140) and (7.142) in configuration space, according to [11]. Parameters, $M = 3, H = 0.3$. Initial condition, $(a_0, s_0, \dot{a}_0, \dot{s}_0) = (1.52, 1.36, 0, 0.62)$

7.5.2   Fully Quantum Case

If H is not very small, it is more appropriate to include κ as a dynamical variable, treating (7.136), (7.139) and (7.140) as a dynamical system for a, κ, and s. In this case, it is not possible to get a closed form solution for the fixed points, which need to be numerically found. For $M = 1, H = 5$, equilibrium is found for $(\kappa, a, s) = (0, 9.15, 3.53)$. Figure 7.10 shows a typical trajectory starting at this initial condition, with $\dot{s}_0 = 0.2$. Under the same parameters but with a smaller initial velocity produces quasi-periodic motion, as shown in Fig. 7.11, where $\dot{s}_0 = 0.05$. Similar simulations shows that for increasing H it becomes more difficult to get regular, quasi-periodic trajectories, pointing for instabilities of quantum nature. In addition, unbounded motion appears for smaller values of the initial velocity.

Further results on the rôle of quantum effects can be obtained in the idealized ultra quantum case where we can neglect all terms in the right-hand sides of (7.136), (7.139) and (7.140) except those containing H^2. In this situation, we get

$$\dot{a} = \frac{12H^2\kappa(1 + 4\kappa^2 a^4)}{a}, \tag{7.151}$$

$$\dot{\kappa} = \frac{3H^2(1 - 16\kappa^4 a^8)}{a^6}, \tag{7.152}$$

$$\ddot{s} = \frac{3H^2}{s^7}, \tag{7.153}$$

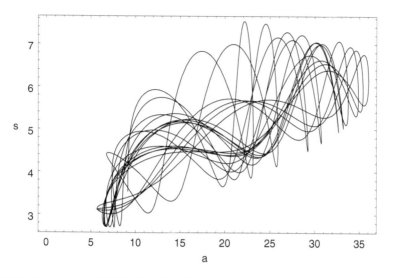

Fig. 7.10 Simulation of (7.136), (7.139) and (7.140) displaying a and s, according to [11]. Parameters: $M = 1, H = 5$. Initial conditions $(\kappa, a, s, \dot{s}) = (0, 9.15, 3.53, 0.20)$

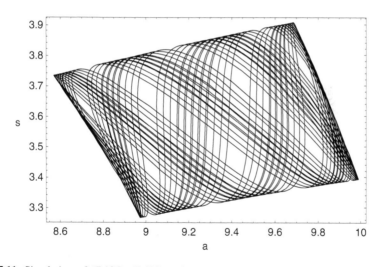

Fig. 7.11 Simulation of (7.136), (7.139) and (7.140) displaying a and s, according to [11]. Parameters: $M = 1, H = 5$. Initial condition at $(\kappa, a, s, \dot{s}) = (0, 9.15, 3.53, 0.05)$

which can be analytically solved yielding

$$a^2 = a_0^2 + \frac{36H^4(t - t_0)^2}{a_0^6}, \quad \kappa^2 = \frac{9a_0^4H^4(t - t_0)^2}{(a_0^8 + 36H^4(t - t_0)^2)^2}, \tag{7.154}$$

where a_0 and t_0 are numerical constants. From (7.154) the conclusion is that the width a of the envelope electric field tends to increase without bound in the ultra quantum limit, while the chirp function κ approaches zero. Similarly, inspection of (7.153) shows that the width $b = s^2$ of the density fluctuation increases without bound, since $\ddot{s} = -\partial V/\partial s$ for a pseudo-potential $V = H^2/2s^6$ having no bound states.

Problems

7.1. Derive the quantum Langmuir dispersion relation in (7.10).

7.2. Repeat the derivation of the quantum Zakharov system using now the equation of state for a isothermal Maxwell–Boltzmann equilibrium, $p = n\kappa_B T$.

7.3. Verify the dispersion relation (7.64) for the quantum four-wave instability.

7.4. Expand (7.69) up to $O(k^6)$. Obtain the wavenumber k_{max} for maximal growth rate. Reproduce Figs. 7.3 and 7.4.

7.5. Check that the Lagrangian density (7.107) yields the one-dimensional quantum Zakharov system.

7.6. Prove the invariance of number of quanta, momentum and energy in (7.119)–(7.121) for the one-dimensional quantum Zakharov system.

7.7. Use the Gaussian *Ansatz* described by (7.105), (7.113), (7.114) and (7.116) to obtain the Lagrangian (7.134) after performing the spatial integration in (7.115).

7.8. Check the conservation law (7.144).

7.9. Repeat the linear stability analysis of (7.140) and (7.148) describing the internal semiclassical oscillations of the quantum Langmuir soliton.

References

1. Ablowitz, M. J. and Segur, H.: Solitons and the Inverse Scattering Transform. SIAM, Philadelphia (1981)
2. Anderson, D.: Variational approach to nonlinear pulse propagation in optical fibers. Phys. Rev. A **27**, 3135–3145 (1983)
3. Davidson, R. C.: Methods in Nonlinear Plasma Theory. Academic Press, New York (1972)
4. Eliasson, B. and Shukla, P. K.: Formation and dynamics of dark solitons and vortices in quantum electron plasmas. Phys. Rev. Lett. **96**, 245001–245005 (2006)
5. Garcia, L. G., Haas, F., Oliveira, L. P. L. and Goedert, J.: Modified Zakharov equations for plasmas with a quantum correction. Phys. Plasmas **12**, 012302–012310 (2005)
6. Gibbons, J., Thornhill, S. G., Wardrop, M. J. and ter Haar, D. H.: On the theory of Langmuir solitons. J. Plasma Phys. **17**, 153–170 (1977)
7. Goldman, M. V.: Strong turbulence of plasma waves. Rev. Mod. Phys. **56**, 709–735 (1984)

8. Gradshteyn, I. S. and Rhyzik, I. M.: Tables of Integrals, Series and Products. Academic Press, New York (1965)
9. Haas, F.: Anisotropic Bose-Einstein condensates and completely integrable dynamical systems. Phys. Rev. A **65**, 33603–33608 (2002)
10. Haas, F., Garcia, L. G., Goedert, J. and Manfredi, G.: Quantum ion-acoustic waves. Phys. Plasmas **10**, 3858–3866 (2003)
11. Haas, F.: Variational approach for the quantum Zakharov system. Phys. Plasmas **14**, 042309–042315 (2007)
12. Malomed, B., Anderson, D., Lisak, M., Quiroga-Teixeiro, M. L. and Stenflo, L.: Dynamics of solitary waves in the Zakharov model equations. Phys. Rev. E **55**, 962–968 (1997)
13. Marklund, M.: Classical and quantum kinetics of the Zakharov system. Phys. Plasmas **12**, 082110–082115 (2005).
14. Nicholson, D. R.: Introduction to Plasma Theory. Wiley, New York (1983)
15. Olver, P.: Applications of Lie Groups to Differential Equations, 2nd ed. Graduate Texts in Mathematics 107. New York, Springer (1993)
16. Sharma, R. P., Batra, K. and Das, S. S.: Variational approach to nonlinear evolution of modulational instability using one-dimensional Zakharov equations. Phys. Plasmas **12**, 092303–092313 (2005)
17. Tang, X. Y. and Shukla, P. K.: Lie symmetry analysis of the quantum Zakharov equations. Phys. Scr. **76**, 665–668 (2007)
18. Thornhill, S. G. and ter Haar, D.: Langmuir turbulence and modulational instability. Phys. Rep. **43**, 43–99 (1978)
19. Zakharov, V. E.: Collapse of Langmuir waves. Sov. Phys. J. Exp. Theor. Phys **35**, 908–914 (1972)

Chapter 8
The Three-Dimensional Quantum Zakharov System

Abstract The results from the last chapter are extended to the three-dimensional situation. First, the three-dimensional quantum Zakharov system is derived from a two-time scales analysis. A Lagrangian formalism and the associated conservation laws are written down. Restricting to the adiabatic and semiclassical case, the system reduces to a quantum vector nonlinear Schrödinger equation (QVNLS) for the envelope electric field. A Lagrangian formalism for this QVNLS equation is used to investigate the behavior of Gaussian shaped solutions (Langmuir wave packets), by means of a time-dependent variational method. Quantum corrections are shown to prevent the collapse of Langmuir wave packets, in both two and three spatial dimensions. The conservation laws of the QVNLS equation are discussed. Finally, we discuss the oscillations of the width of the Langmuir wave packets, as a result from the interplay between classical refraction and quantum diffraction.

8.1 Collapse of Langmuir Wave Packets

In the last chapter, the one-dimensional quantum Zakharov system was investigated in some detail. Quantum effects were shown to be responsible for the suppression of the four-wave decay instability. Also variational methods were employed first for the adiabatic, semiclassical case, which is reducible to a quantum-generalized nonlinear Schrödinger equation for the envelope electric field. Later the general, nonadiabatic, and nonsemiclassical situation was also shown to be amenable to a variational approach. Thanks to the Lagrangian formalism, the spatio-temporal dynamics was replaced by an approximate system of ordinary differential equations, in terms of reasonable trial functions for the density depletion and the envelope electric field. The advantage of the reduced equations is that they are amenable to standard methods of nonlinear analysis, comparing to the original quantum Zakharov equations. For instance, we concluded that quantum effects tend to destroy localized nonlinear structures such as Langmuir solitons, due to tunneling through the self-consistent potential barrier associated with the ponderomotive

F. Haas, *Quantum Plasmas: An Hydrodynamic Approach*, Springer Series on Atomic, Optical, and Plasma Physics 65, DOI 10.1007/978-1-4419-8201-8_8,
© Springer Science+Business Media, LLC 2011

force. The purpose of this chapter is to extend the previous results to the more ambitious case of three spatial dimensions.

In the classical case, it is well known that the solutions for the Zakharov system behaves in a qualitatively different way, according to the dimensionality. For instance, both heuristic arguments and numerical simulations indicate that the ponderomotive force can produce finite-time collapse of Langmuir wave packets in two or three spatial dimensions [7,21,22]. This is in contrast to the one-dimensional case, whose solutions are smooth for all time. A dynamic rescaling method was used for the time-evolution of electrostatic self-similar and asymptotically self-similar solutions in two- and three-dimensions, respectively [10]. Allowing for transverse fields shows that singular solutions of the resulting vector Zakharov equation are weakly anisotropic, for a wide class of initial conditions [14]. The electrostatic nonlinear collapse of Langmuir wave packets in the ionospheric and laboratory plasmas has been observed [4, 16]. Also, the collapse of Langmuir wave packets in beam plasma experiments [3] verifies the basic concepts of strong Langmuir turbulence, as introduced by Zakharov [20]. The analysis of the coupled longitudinal and transverse modes in the classical strong Langmuir turbulence has been less studied [1,2,11], as well as the intrinsically magnetized case [15], which can lead to upper-hybrid wave collapse [18]. Finally, Zakharov-like equations have been proposed for the electromagnetic wave collapse in a radiation background [13].

Our main concern is about the rôle of the quantum diffraction effects against the collapse of localized solutions in the two- and three-dimensional cases. The extra dispersion induced by the Bohm potential term makes the dynamics less violent, a first hint to infer that collapse can be prevented due to quantum effects. To verify the conjecture, a variational procedure will be pursued. Recently, the same question was addressed in a more rigorous way [17], by means of estimates and systematic asymptotic expansions, confirming the qualitative results of the variational approach.

The first task we have is to derive the three-dimensional quantum Zakharov equations. Now, magnetic field perturbations are unavoidable, so that the electromagnetic quantum fluid equations are needed. The general theory on the three-dimensional quantum Zakharov model was presented in [9].

8.2 Derivation of the Three-Dimensional Quantum Zakharov System

For convenience, we write the quantum two-fluid equations in the electromagnetic case. As apparent from Chap. 6, these are

$$\frac{\partial n_e}{\partial t} + \nabla \cdot (n_e \mathbf{u}_e) = 0, \tag{8.1}$$

$$\frac{\partial n_i}{\partial t} + \nabla \cdot (n_i \mathbf{u}_i) = 0, \tag{8.2}$$

$$\frac{\partial \mathbf{u}_e}{\partial t} + \mathbf{u}_e \cdot \nabla \mathbf{u}_e = -\frac{\nabla P_e}{m_e n_e} - \frac{e}{m_e}(\mathbf{E} + \mathbf{u}_e \times \mathbf{B})\frac{\hbar^2}{2m_e^2}\nabla\left(\frac{\nabla^2\sqrt{n_e}}{\sqrt{n_e}}\right), \qquad (8.3)$$

$$\frac{\partial \mathbf{u}_i}{\partial t} + \mathbf{u}_i \cdot \nabla \mathbf{u}_i = \frac{e}{m_i}(\mathbf{E} + \mathbf{u}_e \times \mathbf{B}), \qquad (8.4)$$

as derived after extracting the zeroth- and first-order moments from the electromagnetic Wigner equation. The symbols are the same as in Chap. 6. In contrast to (6.42) and (6.43), no phenomenological dissipation terms are included now in the momentum transport equations, since collisions are not the main concern in what follows. To excellent approximation the ions behave classically due to their large mass, so that no Bohm term was included in the ion force equation. The theory should be adapted to alternative cases, for example, for an electron–positron plasma.

For the electronic fluid pressure, consider the equation of state for a three-dimensional completely degenerate zero-temperature spin $1/2$ gas,

$$P_e = \frac{m_e n_0 v_{Fe}^2}{5}\left(\frac{n_e}{n_0}\right)^3, \qquad (8.5)$$

where all symbols are as before. In a first approximation, the ionic Fermi pressure can be disregarded due to the larger ion mass.

The system (8.1)–(8.4) is coupled to Maxwell's equations,

$$\nabla \cdot \mathbf{E} = \frac{\rho}{\varepsilon_0}, \quad \nabla \cdot \mathbf{B} = 0, \qquad (8.6)$$

$$\nabla \times \mathbf{E} = -\frac{\partial \mathbf{B}}{\partial t}, \quad \nabla \times \mathbf{B} = \mu_0 \mathbf{J} + \mu_0\varepsilon_0\frac{\partial \mathbf{E}}{\partial t}, \qquad (8.7)$$

where the charge and current densities are given, respectively, by

$$\rho = e(n_i - n_e), \quad \mathbf{J} = e(n_i\mathbf{u}_i - n_e\mathbf{u}_e). \qquad (8.8)$$

The model is the same as in the classical plasma case [19, 20], except for the Fermi pressure and Bohm potential in the electron force equation (8.3).

Combining Faraday's and Ampère's laws, we get

$$\mu_0\varepsilon_0\frac{\partial^2 \mathbf{E}}{\partial t^2} + \nabla \times (\nabla \times \mathbf{E}) = -\mu_0\frac{\partial \mathbf{J}}{\partial t}. \qquad (8.9)$$

Due to the larger ion inertia, there are two time scales, with a fast electron and a slow ion dynamics. Hence, it is indicated to introduce a two-time scale decomposition,

$$n_e = n_0 + \delta n_s + \delta n_f, \quad n_i = n_0 + \delta n_s, \qquad (8.10)$$

$$\mathbf{u}_e = \delta\mathbf{u}_s + \delta\mathbf{u}_f, \quad \mathbf{u}_i = \delta\mathbf{u}_s, \qquad (8.11)$$

$$\mathbf{E} = \delta\mathbf{E}_s + \delta\mathbf{E}_f, \quad \mathbf{B} = \delta\mathbf{B}_f. \qquad (8.12)$$

where the subscripts "s" and "f" refer to slowly and rapidly changing quantities, respectively. On averaging over many periods of fast oscillations, we have

$$\langle \delta n_s \rangle = \delta n_s, \quad \langle \delta n_f \rangle = 0, \qquad (8.13)$$

the same applying to the remaining variables.

Following [19], we use the fast component of (8.9) to derive an evolution equation for the electric field. The current density

$$\mathbf{J} = -e \delta n_f \delta \mathbf{u}_s - e(n_0 + \delta n_s + \delta n_f) \delta \mathbf{u}_f \qquad (8.14)$$

on averaging gives

$$\langle \mathbf{J} \rangle = -e \langle \delta n_f \delta \mathbf{u}_f \rangle, \qquad (8.15)$$

so that

$$\mathbf{J}_f = \mathbf{J} - \langle \mathbf{J} \rangle = -e(n_0 + \delta n_s) \delta \mathbf{u}_f - e \delta n_f \delta \mathbf{u}_s - e[\delta n_f \delta \mathbf{u}_f - \langle \delta n_f \delta \mathbf{u}_f \rangle]. \qquad (8.16)$$

The third term on the right-hand side of (8.16) can be neglected using the same estimate as in (7.21), valid for small electric field wavenumbers much smaller than the electron Fermi wavenumber $k_{Fe} = \omega_{pe}/v_{Fe}$. Moreover, the square bracket term in (8.16) can be neglected in terms of a weak turbulence assumption which allows to disregard all harmonic generating terms (see [19] for a detailed justification).

In conclusion, taking into account the different time-scales for the slow and fast quantities, we get

$$\frac{\partial \mathbf{J}_f}{\partial t} = -e(n_0 + \delta n_s) \frac{\partial \delta \mathbf{u}_f}{\partial t}. \qquad (8.17)$$

Hence, from (8.9),

$$\mu_0 \varepsilon_0 \frac{\partial^2 \mathbf{E}_f}{\partial t^2} + \nabla \times (\nabla \times \mathbf{E}_f) = \mu_0 e(n_0 + \delta n_s) \frac{\partial \delta \mathbf{u}_f}{\partial t}. \qquad (8.18)$$

In view of the weak turbulence and small wavenumber assumptions, the fast component of the electron force equation is

$$\frac{\partial \delta \mathbf{u}_f}{\partial t} = \frac{3 \varepsilon_0 v_{Fe}^2}{5 n_0 e} \nabla (\nabla \cdot \delta \mathbf{E}_f) - \frac{e}{m_e} \delta \mathbf{E}_f - \frac{\varepsilon_0 \hbar^2}{4 m_e^2 n_0 e} \nabla \left[\nabla^2 (\nabla \cdot \delta \mathbf{E}_f) \right], \qquad (8.19)$$

where the fast component of Poisson's equation

$$\nabla \cdot \delta \mathbf{E}_f = -\frac{e}{\varepsilon_0} \delta n_f \qquad (8.20)$$

was used to eliminate δn_f. Moreover, all convective terms were neglected [19] and the pressure and Bohm terms linearized. Finally, the magnetic force contribution was discarded since it would contribute a higher-order nonlinearity, $\delta \mathbf{E}_f \gg \delta \mathbf{u}_s \times \delta \mathbf{B}_f$.

Inserting (8.19) into (8.18) and disregarding some higher-order nonlinearities using $n_0 \delta n_f \gg \delta n_s \delta n_f$ the result is

$$\frac{\partial^2 \mathbf{E}_f}{\partial t^2} + \omega_{pe}^2 \delta \mathbf{E}_f + c^2 \nabla \times (\nabla \times \mathbf{E}) - \frac{3}{5} v_{Fe}^2 \nabla (\nabla \cdot \delta \mathbf{E}_f) + \frac{\hbar^2}{4m_e^2} \nabla \left[\nabla^2 (\nabla \cdot \delta \mathbf{E}_f) \right]$$

$$= -\frac{\omega_{pe}^2}{n_0} \delta n_s \delta \mathbf{E}_f, \qquad (8.21)$$

where $c^2 = 1/(\mu_0 \varepsilon_0)$. The left-hand side of (8.21) is depending only on the fast part of the electric field, which is coupled to the slow part of the density perturbation appearing on the right-hand side. Hence, also the equation for δn_s has to be found.

The slow component of the electron force equation is

$$\frac{\partial \delta \mathbf{u}_s}{\partial t} + \delta \mathbf{u}_s \cdot \nabla \delta \mathbf{u}_s + \langle \delta \mathbf{u}_f \cdot \nabla \delta \mathbf{u}_f \rangle + \frac{e}{m_e} \delta \mathbf{E}_s + \frac{e}{m_e} \langle \delta \mathbf{u}_f \times \delta \mathbf{B}_f \rangle$$

$$+ \frac{3}{5} \frac{v_{Fe}^2}{n_0} \nabla \delta n_s - \frac{\hbar^2}{4m_e^2 n_0} \nabla \nabla^2 \delta n_s = 0. \qquad (8.22)$$

To lowest-order on the fast time-scale,

$$\frac{\partial \delta \mathbf{u}_s}{\partial t} = -\frac{e}{m_e} \delta \mathbf{E}_f, \qquad (8.23)$$

which can be combined with the fast part of Faraday's law

$$\nabla \times \delta \mathbf{E}_f = -\frac{\partial \delta \mathbf{E}_f}{\partial t} \qquad (8.24)$$

to obtain

$$\mathbf{B} = \frac{m_e}{e} \nabla \times \delta \mathbf{u}_f. \qquad (8.25)$$

Hence, in (8.22) we get

$$\langle \delta \mathbf{u}_f \cdot \nabla \delta \mathbf{u}_f \rangle + \frac{e}{m_e} \langle \delta \mathbf{u}_f \times \delta \mathbf{B}_f \rangle = \frac{1}{2} \nabla \langle |\delta \mathbf{u}_f|^2 \rangle = \frac{e^2}{2m_e^2 \omega_{pe}^2} \nabla \langle |\delta \mathbf{E}_f|^2 \rangle, \qquad (8.26)$$

the last equality following from

$$\delta \mathbf{u}_f = -ie \delta \mathbf{E}_f / (m_e \omega_{pe}) \qquad (8.27)$$

in a first-order approximation.

Hence, the slow component of the electron force equation becomes

$$\frac{\partial \delta \mathbf{u}_s}{\partial t} + \delta \mathbf{u}_s \cdot \nabla \delta \mathbf{u}_s + \frac{e}{m_e} \delta \mathbf{E}_s + \frac{e^2}{2m_e^2 \omega_{pe}^2} \nabla \left\langle |\delta \mathbf{E}_f|^2 \right\rangle$$

$$+ \frac{3}{5} \frac{v_{Fe}^2}{n_0} \nabla \delta n_s - \frac{\hbar^2}{4m_e^2 n_0} \nabla \nabla^2 \delta n_s = 0. \tag{8.28}$$

In an analogous way, the slow component of the ion force equation is

$$\frac{\partial \delta \mathbf{u}_s}{\partial t} + \delta \mathbf{u}_s \cdot \nabla \delta \mathbf{u}_s - \frac{e}{m_i} \delta \mathbf{E}_s + \frac{e^2}{2m_i^2 \omega_{pe}^2} \nabla \left\langle |\delta \mathbf{E}_f|^2 \right\rangle = 0. \tag{8.29}$$

Eliminating $\delta \mathbf{E}_s$ between (8.28) and (8.29) and using $m_e/m_i \ll 1$ to disregard some terms, we get

$$\frac{\partial \delta \mathbf{u}_s}{\partial t} + \delta \mathbf{u}_s \cdot \nabla \delta \mathbf{u}_s + \frac{e^2}{2m_e m_i \omega_{pe}^2} \nabla \left\langle |\delta \mathbf{E}_f|^2 \right\rangle$$

$$+ \frac{3}{5} \frac{m_e v_{Fe}^2}{m_i n_0} \nabla \delta n_s - \frac{\hbar^2}{4m_e m_i n_0} \nabla \nabla^2 \delta n_s = 0. \tag{8.30}$$

The third term on (8.30) is the so-called ponderomotive force [19], which appears as a radiation pressure force.

The equation of continuity for ions gives

$$\frac{\partial \delta n_s}{\partial t} + \nabla \cdot ((n_0 + \delta n_s) \delta \mathbf{u}_s) = 0, \tag{8.31}$$

which, combined with (8.30) and linearizing with respect to δn_s and $\delta \mathbf{u}_s$ gives

$$\frac{\partial^2}{\partial t^2} \left(\frac{\delta n_s}{n_0} \right) - \frac{3}{5} c_s^2 \nabla^2 \left(\frac{\delta n_s}{n_0} \right) = \frac{\varepsilon_0}{2m_i n_0} \nabla^2 \left\langle |\delta \mathbf{E}_f|^2 \right\rangle - \frac{\hbar^2}{4m_e m_i n_0} \nabla^4 \left(\frac{\delta n_s}{n_0} \right), \tag{8.32}$$

where

$$c_s = \left(\frac{m_e v_{Fe}^2}{m_i} \right)^{1/2} \tag{8.33}$$

is the ion-acoustic velocity in the case of a degenerate Fermi electron gas.

Equations (8.21) and (8.32) form a closed system for δn_s and $\delta \mathbf{E}_f$. However, some improvement can be obtained defining the slowly varying envelope electric field $\tilde{\mathbf{E}}$ via

$$\delta \mathbf{E}_f = \frac{1}{2} \left(\tilde{\mathbf{E}} e^{-i\omega_{pe}t} + \tilde{\mathbf{E}}^* e^{i\omega_{pe}t} \right). \tag{8.34}$$

Disregarding the second-order time-derivative of the slowly varying envelope electric field, and using $\langle|\delta \mathbf{E}_f|^2\rangle = (1/2)\langle|\tilde{\mathbf{E}}_f|^2\rangle$, we finally obtain the quantum corrected 3D Zakharov equations,

$$i\omega_{pe}\frac{\partial \tilde{\mathbf{E}}}{\partial t} - c^2 \nabla \times (\nabla \times \tilde{\mathbf{E}}) + \frac{3}{5}v_{Fe}^2\nabla(\nabla \cdot \tilde{\mathbf{E}}) - \frac{\hbar^2}{4m_e^2}\nabla\left[\nabla^2(\nabla \cdot \tilde{\mathbf{E}})\right] = \frac{\delta n_s}{n_0}\omega_{pe}^2\tilde{\mathbf{E}},$$

$$(8.35)$$

$$\frac{\partial^2}{\partial t^2}\left(\frac{\delta n_s}{n_0}\right) - \frac{3}{5}c_s^2\nabla^2\left(\frac{\delta n_s}{n_0}\right) + \frac{\hbar^2}{4m_e m_i n_0}\nabla^4\left(\frac{\delta n_s}{n_0}\right) = \frac{\varepsilon_0}{4m_i n_0}\nabla^2\langle|\tilde{\mathbf{E}}|^2\rangle.$$

$$(8.36)$$

In comparison to the classical Zakharov system (see (2.48a) and (2.48b) of [19]), there is the inclusion of the extra dispersive terms proportional to $\hbar^2$ in (8.35) and (8.36). Other quantum difference is the presence of the Fermi speed instead of the thermal speed in the last term at the left-hand side of (8.35). From the qualitative point of view, the terms proportional to $\hbar^2$ are responsible for extra dispersion which can avoid collapsing of Langmuir envelopes, at least in principle. Finally, notice the nontrivial form of the fourth-order derivative term in (8.35). It is not simply proportional to $\nabla^4\tilde{\mathbf{E}}$ as could be wrongly guessed from the quantum Zakharov equations in $1+1$ dimensions, where there is a $\sim \partial^4\tilde{\mathbf{E}}/\partial x^4$ contribution [5].

It is useful to consider the rescaling

$$\bar{\mathbf{r}} = \frac{2\sqrt{5\mu/3}\,\omega_{pe}\,\mathbf{r}}{v_{Fe}}, \quad \bar{t} = 2\mu\,\omega_{pe}t,$$

$$n = \frac{\delta n_s}{4\mu n_0}, \quad \mathscr{E} = \frac{e\tilde{\mathbf{E}}}{4\sqrt{3\mu/5}\,m_e\omega_{pe}v_{Fe}}, \qquad (8.37)$$

where $\mu = m_e/m_i$. Then, dropping the bars in $\mathbf{r}, t$, we obtain

$$i\frac{\partial \mathscr{E}}{\partial t} - \frac{5c^2}{3v_{Fe}^2}\nabla \times (\nabla \times \mathscr{E}) + \nabla(\nabla \cdot \mathscr{E}) = n\mathscr{E} + H\nabla\left[\nabla^2(\nabla \cdot \mathscr{E})\right], \qquad (8.38)$$

$$\frac{\partial^2 n}{\partial t^2} - \nabla^2 n - \nabla^2(|\mathscr{E}|^2) + H\nabla^4 n = 0, \qquad (8.39)$$

where

$$H = \frac{m_e}{m_i}\left(\frac{5\hbar\,\omega_{pe}}{3\kappa_B T_{Fe}}\right)^2 \qquad (8.40)$$

is a nondimensional parameter associated with the quantum effects, in terms of $\kappa_B T_{Fe} = m_e v_{Fe}^2$. Usually, it is a very small quantity, but it is nevertheless interesting to retain the $\sim H$ terms, specially for the collapse scenarios. The reason is not only in a general theoretical motivation, but also because from some simple estimates one concludes that these terms become of the same order as some of other terms in (8.35) and (8.36) provided that the characteristic length L for the spatial derivatives becomes as small as the mean inter-particle distance, $L \sim n_0^{-1/3}$. Precisely, in a collapsing scenario one would have large spatial gradients and then the quantum effects would come into play. However, of course the quantum Zakharov equations are not able to describe the late stages of the collapse, since they do not include dissipation, which is unavoidable for short scales. In the left-hand side of (8.38), the $\nabla(\nabla \cdot \mathscr{E})$ term is retained because the $\sim c^2/v_{Fe}^2$ transverse term disappears in the electrostatic approximation.

In the adiabatic limit, neglecting $\partial^2 n/\partial t^2$ in (8.39) and under appropriated boundary conditions, it follows that

$$n = -|\mathscr{E}|^2 + H\nabla^2 n, \tag{8.41}$$

It is not so easy to directly express n as a function of $|\mathscr{E}|$ as in the classical case. Therefore, the adiabatic limit is not enough to derive a vector nonlinear Schrödinger equation.

8.3   Lagrangian Structure and Conservation Laws

The quantum Zakharov equations (8.38) and (8.39) can be described by the Lagrangian density

$$\mathscr{L} = \frac{i}{2}\left(\mathscr{E}^* \cdot \frac{\partial \mathscr{E}}{\partial t} - \mathscr{E} \cdot \frac{\partial \mathscr{E}^*}{\partial t}\right) - \frac{5c^2}{3v_{Fe}^2}|\nabla \times \mathscr{E}|^2 - |\nabla \cdot \mathscr{E}|^2 - H|\nabla(\nabla \cdot \mathscr{E})|^2$$
$$+ n\left(\frac{\partial \alpha}{\partial t} - |\mathscr{E}|^2\right) - \frac{1}{2}\left(n^2 + H|\nabla n|^2 + |\nabla \alpha|^2\right), \tag{8.42}$$

where n, the auxiliary function α and the components of $\mathscr{E}, \mathscr{E}^*$ are regarded as independent fields.

Remark. For the particular form (8.42) and for a generic field ψ, one computes the functional derivative as

$$\frac{\delta \mathscr{L}}{\delta \psi} = \frac{\partial \mathscr{L}}{\partial \psi} - \frac{\partial}{\partial r_i}\frac{\partial \mathscr{L}}{\partial \psi/\partial r_i} - \frac{\partial}{\partial t}\frac{\partial \mathscr{L}}{\partial \psi/\partial t} + \frac{\partial^2}{\partial r_i \partial r_j}\frac{\partial \mathscr{L}}{\partial^2 \psi/\partial r_i \partial r_j}, \tag{8.43}$$

using the summation convention and where r_i are Cartesian components.

Taking the functional derivatives with respect to n and α, we have

$$\frac{\partial \alpha}{\partial t} = n + |\mathscr{E}|^2 - H\nabla^2 n, \tag{8.44}$$

and

$$\frac{\partial n}{\partial t} = \nabla^2 \alpha, \tag{8.45}$$

respectively. Eliminating α from (8.44) and (8.45), we obtain the low frequency equation. In addition, the functional derivatives with respect to $\mathscr{E}^*$ and $\mathscr{E}$ produce the high-frequency equation and its complex conjugate. The present formalism is inspired by the Lagrangian formulation of the classical Zakharov equations [6].

The quantum Zakharov equations admit as exact conserved quantities the "number of plasmons" of the Langmuir field,

$$N = \int |\mathscr{E}|^2 \, d\mathbf{r}, \tag{8.46}$$

the linear momentum (with components P_i, $i = x, y, z$),

$$P_i = \int \left[\frac{i}{2} \left(\mathscr{E}_j \frac{\partial \mathscr{E}_j^*}{\partial r_i} - \mathscr{E}_j^* \frac{\partial \mathscr{E}_j}{\partial r_i} \right) - n \frac{\partial \alpha}{\partial r_i} \right] d\mathbf{r} \tag{8.47}$$

and the Hamiltonian,

$$\mathscr{H} = \int \left[n|\mathscr{E}|^2 + \frac{5c^2}{3v_{\mathrm{Fe}}^2} |\nabla \times \mathscr{E}|^2 + |\nabla \cdot \mathscr{E}|^2 + H |\nabla(\nabla \cdot \mathscr{E})|^2 \right.$$
$$\left. + \frac{1}{2} \left(n^2 + H|\nabla n|^2 + |\nabla \alpha|^2 \right) \right] d\mathbf{r}. \tag{8.48}$$

Furthermore, there is also a preserved angular momenta functional, but it is not relevant in the present work. These four conserved quantities can be associated, through Noether's theorem, to the invariance of the action under gauge transformation, time translation, space translation and rotations, respectively. The conservation laws can be used, for example, to test the accuracy of numerical procedures. Also, observe that equations (8.39) and (8.41) for the adiabatic limit are described by the same Lagrangian density (8.42). In this approximation, it suffices to set $\alpha \equiv 0$.

In addition to the adiabatic limit, (8.41) can be further approximated to

$$n = -|\mathscr{E}|^2 - H\nabla^2(|\mathscr{E}|^2), \tag{8.49}$$

assuming that the quantum term is a perturbation. In this way and using (8.38), a quantum vector nonlinear Schrödinger equation (QVNLS) is derived

$$i\frac{\partial \mathscr{E}}{\partial t} + \nabla(\nabla \cdot \mathscr{E}) - \frac{5c^2}{3v_{Fe}^2} \nabla \times (\nabla \times \mathscr{E}) + |\mathscr{E}|^2 \mathscr{E}$$

$$= H\nabla\left[\nabla^2(\nabla \cdot \mathscr{E})\right] - H\mathscr{E}\nabla^2(|\mathscr{E}|^2). \tag{8.50}$$

The appropriate Lagrangian density $\mathscr{L}_{ad,sc}$ for the semiclassical equation (8.50) is given by

$$\mathscr{L}_{ad,sc} = \frac{i}{2}\left(\mathscr{E}^* \cdot \frac{\partial \mathscr{E}}{\partial t} - \mathscr{E} \cdot \frac{\partial \mathscr{E}^*}{\partial t}\right) - \frac{5c^2}{3v_{Fe}^2}|\nabla \times \mathscr{E}|^2 - |\nabla \cdot \mathscr{E}|^2$$

$$-H|\nabla(\nabla \cdot \mathscr{E})|^2 + \frac{1}{2}|\mathscr{E}|^4 - \frac{H}{2}\left|\nabla[|\mathscr{E}|^2]\right|^2, \tag{8.51}$$

where the independent fields are taken as $\mathscr{E}$ and $\mathscr{E}^*$ components.

The expression N for the number of plasmons in (8.46) remain valid as a constant of motion in the joint adiabatic and semiclassical limit, as well as the momentum $\mathbf{P}$ in (8.47) with $\alpha \equiv 0$. Finally, the Hamiltonian

$$\mathscr{H}_{ad,sc} = \int\left[\frac{5c^2}{3v_{Fe}^2}|\nabla \times \mathscr{E}|^2 + |\nabla \cdot \mathscr{E}|^2 + H|\nabla(\nabla \cdot \mathscr{E})|^2 - \frac{1}{2}|\mathscr{E}|^4 + \frac{H}{2}\left|\nabla[|\mathscr{E}|^2]\right|^2\right]d\mathbf{r}$$

$$\tag{8.52}$$

is also a conserved quantity.

In the following, the influence of the quantum terms in the right-hand side of (8.50) are investigated, assuming adiabatic conditions for collapsing quantum Langmuir envelopes. Other scenarios for collapse, like the supersonic one [10, 14], could also be relevant.

8.4   Variational Solution in Two Dimensions

Consider the adiabatic semiclassical system described by (8.50). We refer to localized solution for this QVNLS equation as quantum "Langmuir wave packets," or envelopes. As discussed in detail in [6] in the purely classical case, Langmuir wave packets will become singular in a finite time, provided the energy is not bounded from below. Of course, explicit analytic three-dimensional Langmuir envelopes are difficult to derive. A fruitful approach is to make use of the Lagrangian structure for deriving approximate solutions. This approach has been pursued in [12] for the classical and in [8] for the quantum Zakharov system. Both studies considered the internal vibrations of Langmuir envelopes in one spatial dimension.

Presently, we shall apply the time-dependent variational method for the higher-dimensional cases. A priori, it is expected that the quantum corrections would inhibit the collapse of localized solutions, in view of wave packet spreading. To check this conjecture, and to have more definite information on the influence of the quantum terms, first we consider the following *Ansatz*,

$$\mathscr{E} = \left(\frac{N}{\pi}\right)^{1/2} \frac{1}{\sigma} \exp\left(-\frac{\rho^2}{2\sigma^2}\right) \exp\left(i(\Theta + k\rho^2)\right) (\cos\phi, \sin\phi, 0), \quad (8.53)$$

which is appropriate for two spatial dimensions. Here, σ, k, Θ, and ϕ are real functions of time, and $\rho = \sqrt{x^2 + y^2}$. The normalization condition (8.46) is automatically satisfied (in 2D the spatial integrations reduce to integrations on the plane). Other localized forms, involving, for example, a sech-type dependence, could have been also proposed. Here, a Gaussian form was suggested mainly for the sake of simplicity. Notice that the envelope electric field (8.53) is not necessarily electrostatic: it can carry a transverse $(\nabla \times \mathscr{E} \neq 0)$ component.

The free functions in (8.53) should be determined by extremization of the action functional associated with the Lagrangian density (8.51). A straightforward calculation gives

$$L_2 \equiv \int \mathscr{L}_{\text{ad,sc}} \, dx \, dy = -N \left[\dot{\Theta} + \sigma^2 k + \frac{10c^2}{3v_{\text{Fe}}^2} k^2 \sigma^2 + \frac{1}{2}\left(\frac{5c^2}{3v_{\text{Fe}}^2} - \frac{N}{2\pi}\right)\frac{1}{\sigma^2} \right.$$

$$\left. + 8Hk^2 + 16Hk^4\sigma^4 + \left(1 + \frac{N}{2\pi}\right)\frac{H}{\sigma^4} \right], \quad (8.54)$$

where only the main quantum contributions are retained. Now L_2 is the Lagrangian for a finite-dimensional mechanical system, since the spatial dependence of the envelope electric field was defined in advance via (8.53). Of special interest is the behavior of the width function σ. For a collapsing solution one could expect that σ goes to zero in a finite time. The phase Θ and the chirp function k should be regarded as auxiliary fields. Notice that L_2 is not dependent on the angle ϕ, which remains arbitrary as far as the variational method is concerned.

Applying the functional derivative of L_2 with respect to Θ, we obtain

$$\frac{\delta L_2}{\delta \Theta} = 0 \quad \rightarrow \quad \dot{N} = 0, \quad (8.55)$$

so that the variational solution preserves the number of plasmons, as expected. The remaining Euler–Lagrange equations are

$$\frac{\delta L_2}{\delta k} = 0 \quad \rightarrow \quad \sigma\dot{\sigma} = \frac{10c^2}{3v_{\text{Fe}}^2}\sigma^2 k + 8Hk + 32H\sigma^4 k^3, \quad (8.56)$$

$$\frac{\delta L_2}{\delta \sigma} = 0 \;\rightarrow\; \sigma \dot{k} = -\frac{10c^2}{3v_{\text{Fe}}^2} k^2 \sigma + \frac{1}{2}\left(\frac{5c^2}{3v_{\text{Fe}}^2} - \frac{N}{2\pi}\right)\frac{1}{\sigma^3} - 32Hk^4\sigma^3$$
$$+\left(1 + \frac{N}{2\pi}\right)\frac{2H}{\sigma^5}. \tag{8.57}$$

The exact solution of the nonlinear system (8.56) and (8.57) is difficult to obtain, but at least the dynamics was reduced to a set of ordinary differential equations.

It is instructive to analyze the purely classical ($H \equiv 0$) case first. This is specially true, since to our knowledge the time-dependent variational method was not applied to the QVNLS equation (8.50), even for classical systems. The reason can be due to the algebraic complexity induced by the transverse term.

When $H = 0$, (8.56) gives

$$k = \frac{3v_{\text{Fe}}^2 \dot{\sigma}}{10c^2\sigma}. \tag{8.58}$$

Inserting this result in (8.57), we have

$$\ddot{\sigma} = -\frac{\partial V_{2c}}{\partial \sigma}, \tag{8.59}$$

where the classical pseudo-potential V_{2c} is

$$V_{2c} = \frac{5c^2}{6v_{\text{Fe}}^2}\left(\frac{5c^2}{3v_{\text{Fe}}^2} - \frac{N}{2\pi}\right)\frac{1}{\sigma^2}. \tag{8.60}$$

From (8.60) it is evident that the repulsive character of the pseudo-potential will be converted into an attractive one, whenever the number of plasmons exceeds a threshold,

$$N > \frac{10\pi c^2}{3v_{\text{Fe}}^2}, \tag{8.61}$$

a condition for Langmuir wave packet collapse in the classical two-dimensional case. The interpretation of the result is as follows. When the number of plasmons satisfy (8.61), the refractive $\sim |\mathscr{E}|^4$ term dominates over the dispersive terms in the Lagrangian density (8.51), producing a singularity in a finite time. Finally, notice the ballistic motion when $N = 10\pi c^2/(3v_{\text{Fe}}^2)$, which can also lead to singularity.

Further insight follows after evaluating the energy integral (8.52) with the *Ansatz* (8.53), which gives, after eliminating k,

$$\mathscr{H}_{\text{ad,sc}} = \frac{3Nv_{\text{Fe}}^2}{5c^2}\left[\frac{\dot{\sigma}^2}{2} + V_{2c}\right] \quad (H = 0, \quad 2D). \tag{8.62}$$

Of course, this energy first integral could be directly obtained from (8.59). However, the plausibility of the variational solution is reinforced, since (8.62) shows that it preserves the exact constant of motion $\mathcal{H}_{\text{ad,sc}}$. In addition, in the attractive (collapsing) case the energy (8.62) is not bounded from below.

In the quantum ($H \neq 0$) case, (8.56) becomes a cubic equation in k, whose exact solution is too cumbersome to be of practical use. It is better to proceed by successive approximations, taking into account that the quantum and electromagnetic terms are small. In this way, one arrives at

$$\ddot{\sigma} = -\frac{\partial V_2}{\partial \sigma}, \tag{8.63}$$

where the pseudo-potential V_2 is

$$V_2 = \frac{5c^2}{6v_{\text{Fe}}^2} \left(\frac{5c^2}{3v_{\text{Fe}}^2} - \frac{N}{2\pi} \right) \frac{1}{\sigma^2} + \frac{5Hc^2}{3v_{\text{Fe}}^2} \left(1 + \frac{N}{2\pi} \right) \frac{1}{\sigma^4}. \tag{8.64}$$

Now, even if the threshold (8.61) is exceeded, the repulsive $\sim \sigma^{-4}$ quantum term in V_2 will prevent singularities. This adds quantum diffraction as another physical mechanism, besides dissipation and Landau damping, so that collapsing Langmuir wave packets are avoided in the QVNLS equation. Also, similarly to (8.62), it can be shown that the approximate dynamics preserves the energy integral, even in the quantum case. Indeed, calculating from (8.52) and the variational solution gives $\mathcal{H}_{\text{ad,sc}}$ as

$$\mathcal{H}_{\text{ad,sc}} = \frac{3Nv_{\text{Fe}}^2}{5c^2} \left[\frac{\dot{\sigma}^2}{2} + V_2 \right] \quad (H \geq 0, \quad 2D). \tag{8.65}$$

From (8.63), obviously $\mathcal{H}_{\text{ad,sc},2} = 0$.

It should be noticed that oscillations of purely quantum nature are obtained when the number of plasmons exceeds the threshold (8.61). Indeed, in this case the pseudo-potential V_2 in (8.64) assumes a potential well form as shown in Fig. 8.1, which clearly admits oscillations around a minimum $\sigma = \sigma_m$. Here,

$$\sigma_m = 2 \left[\frac{H(1 + N/2\pi)}{N/2\pi - 5c^2/(3v_{\text{Fe}}^2)} \right]^{1/2}. \tag{8.66}$$

At this place, one get the minimum value of V_2,

$$V_2(\sigma_m) = -\frac{5c^2}{48H v_{\text{Fe}}^2} \frac{(N/2\pi - 5c^2/(3v_{\text{Fe}}^2))^2}{1 + N/2\pi} > -\frac{1}{16H} \left(\frac{N}{2\pi} - \frac{5c^2}{3v_{\text{Fe}}^2} \right)^2, \tag{8.67}$$

the last inequality follows since (8.61) is assumed. Therefore, a deepest potential well is obtained when N is increasing. Also, for too large quantum effects the trapping of the localized electric field in this potential well would be difficult, since $V_2(\sigma_m) \to 0_-$ as H increases. This is due to the dispersive nature of the quantum corrections.

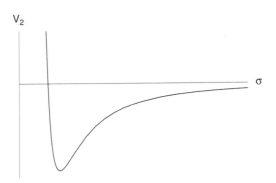

Fig. 8.1 The qualitative form of the pseudo-potential in (8.64) for $N > 10\pi c^2/(3v_{\mathrm{Fe}}^2)$

The frequency ω of the small amplitude oscillations is derived linearizing (8.63) around the equilibrium point (8.66). Restoring physical coordinates via (8.37) this frequency is calculated as

$$\omega = \sqrt{\frac{5}{6}\frac{c}{v_{\mathrm{Fe}}}}\left(\frac{3\kappa_B T_{\mathrm{Fe}}}{5\hbar\omega_{\mathrm{pe}}}\right)^2 \frac{(N/2\pi - 5c^2/(3v_{\mathrm{Fe}}^2))^{3/2}}{1 + N/2\pi}\,\omega_{\mathrm{pe}}$$

$$< \sqrt{\frac{5}{6}\frac{v_{\mathrm{Fe}}}{c}}\left(\frac{3\kappa_B T_{\mathrm{Fe}}}{5\hbar\omega_{\mathrm{pe}}}\right)^2\left(\frac{N}{2\pi} - \frac{5c^2}{3v_{\mathrm{Fe}}^2}\right)^{3/2}\omega_{\mathrm{pe}}. \tag{8.68}$$

To conclude, the variational solution suggests that the extra dispersion arising from the quantum terms would inhibit the collapse of Langmuir wave packets in two spatial dimensions. Moreover, for sufficient electric field energy (which is proportional to N), instead of collapse there will be oscillations of the width of the localized solution, due to the competition between classical refraction and quantum diffraction. The frequency of linear oscillations is then given by (8.68). The emergence of a pulsating Langmuir envelope is a qualitatively new phenomena, which may be tested quantitatively in experiments.

8.5 Variational Solution in Three Dimensions

It is worth to study the dynamics of localized solutions for the QVNLS equation (8.50) in fully three-dimensional space. For this purpose, we consider the Gaussian form

$$\mathcal{E} = \left(\frac{N}{(\sqrt{\pi}\sigma)^3}\right)^{1/2}\exp\left[-\frac{r^2}{2\sigma^2} + i(\Theta + kr^2)\right](\cos\phi\sin\theta, \sin\phi\sin\theta, \cos\theta),$$
$$\tag{8.69}$$

where $\sigma, k, \Theta, \theta,$ and ϕ are real functions of time and $r = \sqrt{x^2 + y^2 + z^2}$, applying the time-dependent variational method just like in the last section. The normalization condition (8.46) is automatically satisfied with (8.69), which can also support a transverse ($\nabla \times \mathscr{E} \neq 0$) part.

Proceeding as before, the Lagrangian

$$L_3 \equiv \int \mathscr{L}_{\mathrm{ad,sc}} \, \mathbf{dr} = -N \left[\dot{\Theta} + \frac{3}{2}\sigma^2 k + \frac{12c^2}{5v_{\mathrm{Fe}}^2}k^2\sigma^2 + \frac{3c^2}{5v_{\mathrm{Fe}}^2\sigma^2} - \frac{N}{4\sqrt{2}\pi^{3/2}\sigma^3} \right.$$

$$\left. + 10Hk^2 + 20Hk^4\sigma^4 + \frac{5H}{4\sigma^4} + \frac{3HN}{4\sqrt{2}\pi^{3/2}\sigma^5} \right] \qquad (8.70)$$

is derived. In comparison to the reduced 2D-Lagrangian in (8.54), there are different numerical factors as well as qualitative changes due to higher-order nonlinearities. Also, the angular variables θ and ϕ do not appear in L_3. Moreover, the algebra is more complicated and a symbolic computation package is advisable to obtain (8.70).

The main remaining task is to analyze the dynamics of the width σ as a function of time. This is achieved from the Euler–Lagrange equations for the action functional associated with L_3. As before, $\delta L_3 / \delta \Theta = 0$ gives $\dot{N} = 0$, a consistency test satisfied by the variational solution. The other functional derivatives yield

$$\frac{\delta L_3}{\delta k} = 0 \rightarrow \sigma\dot{\sigma} = \frac{4k}{3}\left[\frac{6c^2}{5v_{\mathrm{Fe}}^2}\sigma^2 + 5H(1 + 4k^2\sigma^4) \right], \qquad (8.71)$$

$$\frac{\delta L_3}{\delta \sigma} = 0 \rightarrow \sigma\dot{k} = \frac{1}{3}\left[-\frac{24c^2}{5v_{\mathrm{Fe}}^2}k^2\sigma + \frac{6c^2}{5v_{\mathrm{Fe}}^2\sigma^3} - \frac{3N}{4\sqrt{2}\pi^{3/2}\sigma^4} \right.$$

$$\left. - 80Hk^4\sigma^3 + \frac{5H}{\sigma^5} + \frac{15HN}{4\sqrt{2}\pi^{3/2}\sigma^6} \right]. \qquad (8.72)$$

In the formal classical limit ($H \equiv 0$), and using (8.71) to eliminate k, we obtain

$$\ddot{\sigma} = -\frac{\partial V_{3c}}{\partial \sigma}, \qquad (8.73)$$

where now the pseudo-potential V_{3c} is

$$V_{3c} = \frac{3c^2}{5v_{\mathrm{Fe}}^2}\left(\frac{8c^2}{15v_{\mathrm{Fe}}^2\sigma^2} - \frac{2N}{9\sqrt{2}\pi^{3/2}\sigma^3} \right). \qquad (8.74)$$

The form (8.74) shows a generic singular behavior, since the attractive $\sim\sigma^{-3}$ term will dominate for sufficiently small σ, irrespective of the value of N. Hence, in fully

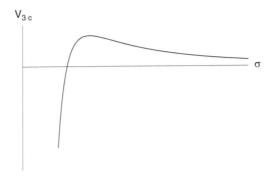

Fig. 8.2 The qualitative form of the pseudo-potential V_{3c} in (8.74)

three-dimensional space there is more "room" for a collapsing dynamics. Figure 8.2 shows the qualitative form of V_{3c}, attaining a maximum at $\sigma = \sigma_M$, where

$$\sigma_M = \frac{5\, v_F^2\, N}{8\sqrt{2}\,\pi^{3/2} c^2}. \tag{8.75}$$

By (8.72) and using successive approximations in the parameter H to eliminate k via (8.71), we obtain

$$\ddot{\sigma} = -\frac{\partial V_3}{\partial \sigma}, \tag{8.76}$$

where

$$V_3 = \frac{8 c^2}{5 v_{Fe}^2}\left[\frac{c^2}{5 v_{Fe}^2\, \sigma^2} - \frac{N}{12\sqrt{2}\,\pi^{3/2}\sigma^3} + \frac{5H}{12\sigma^4} + \frac{HN}{4\sqrt{2}\,\pi^{3/2}\sigma^5}\right]. \tag{8.77}$$

The quantum terms are repulsive and prevent collapse, since they dominate for sufficiently small σ. Moreover, when $H \neq 0$ an oscillatory behavior is possible, provided a certain condition, to be explained in the following, is meet.

To examine the possibility of oscillations, consider $V_3'(\sigma) = 0$, the equation for the critical points of V_3. Under the rescaling $s = \sigma/\sigma_M$, where σ_M (defined in (8.75)) is the maximum of the purely classical pseudo-potential, the equation for the critical points read

$$V_3' = 0 \;\rightarrow\; s^3 - s^2 + \frac{4g}{27} = 0, \tag{8.78}$$

where

$$g = \frac{864\,\pi^3\, H c^4}{5 N^2\, v_{Fe}^4} \tag{8.79}$$

is a new dimensionless parameter. In deriving (8.78), it was omitted a term negligible except if $s \sim c^2/v_{Fe}^2$, which is unlikely.

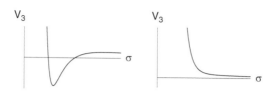

Fig. 8.3 The qualitative form of the pseudo-potential V_3 in (8.77) for $g < 1$ (on the *left*) and $g > 1$ (on the *right*)

The quantity g plays a decisive rôle on the shape of V_3. Indeed, calculating the discriminant shows that the solutions to the cubic in (8.78) are as follows: (a) $g < 1 \rightarrow$ three distinct real roots (one negative and two positive); (b) $g = 1 \rightarrow$ one negative root, one (positive) double root; (c) $g > 1 \rightarrow$ one (negative) real root, two complex conjugate roots. Therefore, $g < 1$ is the condition for the existence of a potential well, which can support oscillations. This is shown in Fig. 8.3. The analytic solutions of the cubic in (8.78) are cumbersome and will be omitted.

Restoring physical coordinates, the necessary condition for oscillations is rewritten as

$$g < 1 \quad \rightarrow \quad \frac{\varepsilon_0}{2} \int |\tilde{\mathbf{E}}|^2 \, d\mathbf{r} > \frac{\sqrt{50\pi}}{\Gamma} m_e v_{Fe} c, \tag{8.80}$$

where $\Gamma = e^2/4\pi\varepsilon_0 \hbar c \simeq 1/137$ is the fine structure constant. From (8.80) it is seen that for sufficient electrostatic energy the width σ of the localized envelope field can show oscillations, supported by the competition between classical refraction and quantum diffraction. Also, due to the Fermi pressure, for large particle densities the inequality (8.80) becomes more difficult to be met, since $v_{Fe} \sim n_0^{1/3}$. For example, when $n_0 \sim 10^{36}$ m^{-3} (white dwarf), the right-hand side of (8.80) is 0.8 GeV. For $n_0 \sim 10^{33}$ m^{-3} (the next generation intense laser-solid density plasma experiments), it is 74.2 MeV.

Finally, notice that $\mathcal{H}_{\mathrm{ad,sc}}$ from (8.52), evaluated with the variational solution (8.69), is proportional to $\dot{\sigma}^2/2 + V_3$, which is a constant of motion for (8.76). Therefore, the approximate solution preserves one of the basic first integrals of the QVNLS equation (8.50), as it should.

Problems

8.1. Rederive the three-dimensional quantum Zakharov system using a general polytropic equation of state, $P_e = P_0(n/n_0)^\gamma$, where P_0 and γ are constants.

8.2. Show that the quantum Zakharov equations (8.38) and (8.39) can be described by the Lagrangian density (8.42).

8.3. Using the quantum Zakharov system, directly check the invariance of the quantities in (8.46)–(8.48), associated with the plasmon number, momentum and energy conservation.

8.4. Verify that (8.51) provides a Lagrangian density for the QVNLS equation.

8.5. Obtain the Lagrangian L_2 in (8.54) for the two-dimensional time-dependent variational solution for the QVNLS equation.

8.6. Check (8.68) for the frequency of small amplitude width oscillations in the two-dimensional case.

8.7. Obtain the Lagrangian L_3 in (8.70) for the three-dimensional time-dependent variational solution for the QVNLS equation. Using a symbolic calculus package may be a good idea.

8.8. Consider the potential V_3 in (8.77) for the three-dimensional wave packet dynamics. Linearize it and derive the frequency of small amplitude oscillations, in the periodic motion case.

8.9. Calculate the energy in (8.52) for the three-dimensional time-dependent variational solution. Show that it is proportional to the reduced, one-dimensional Hamiltonian $\dot{\sigma}^2/2 + V_3$ for the width dynamics.

References

1. Akimoto, K., Rowland, H. L. and Papadopoulos, K.: Electromagnetic-radiation from strong Langmuir turbulence. Phys. Fluids **31**, 2185 (1988)
2. Alinejad, H., Robinson, P. A., Cairns, I. H., Skjaeraasen, O. and Sobhanian, C.: Structure of Langmuir and electromagnetic collapsing wave packets in two-dimensional strong plasma turbulence. Phys. Plasmas **14**, 082304–082314 (2007)
3. Cheung, P. Y. and Wong, A. Y.: Nonlinear evolution of electron-beam plasma interactions. Phys. Fluids **18**, 1538–1548 (1985)
4. Dubois, D. F., Hanssen, A., Rose, H. A. and Russel, D.: Space and time distribution of HF excited Langmuir turbulence in the ionosphere – comparison of theory and experiment. J. Geophys. Res. **98**, 17543–17567 (1993)
5. Garcia, L. G., Haas, F., Oliveira, L. P. L. and Goedert, J.: Modified Zakharov equations for plasmas with a quantum correction. Phys. Plasmas **12**, 012302–012310 (2005)
6. Gibbons, J., Thornhill, S. G., Wardrop, M. J. and ter Haar, D. H.: On the theory of Langmuir solitons. J. Plasma Phys. **17**, 153–170 (1977)
7. Goldman, M. V.: Strong turbulence of plasma waves. Rev. Mod. Phys. **56**, 709–735 (1984)
8. Haas, F.: Variational approach for the quantum Zakharov system. Phys. Plasmas **14**, 042309–042315 (2007)
9. Haas, F. and Shukla, P. K.: Quantum and classical dynamics of Langmuir wave packets. Phys. Rev. **79**, 066402–066501 (2009)
10. Landman, M., Papanicolaou, G. C., Sulem, C., Sulem, P. L. and Wang, X. P.: Stability of isotropic self-similar dynamics for scalar-wave collapse. Phys. Rev. A **46**, 7869–7876 (1992)
11. Li, L. H. and Li, X. Q.: Three-wave interactions in strong Langmuir turbulence. Phys. Fluids B **5**, 3819–3822 (1993)

12. Malomed, B., Anderson, D., Lisak, M., Quiroga-Teixeiro, M. L. and Stenflo, L.: Dynamics of solitary waves in the Zakharov model equations. Phys. Rev. E **55**, 962–968 (1997)
13. Marklund, M., Brodin, G. and Stenflo, L.: Electromagnetic wave collapse in a radiation background. Phys. Rev. Lett. **91**, 163601–163605 (2003)
14. Papanicolaou, G. C., Sulem, C., Sulem, P. L. and Wang, X. P.: Singular solutions of the Zakharov equations for Langmuir turbulence. Phys. Fluids B **3**, 969–980 (1991)
15. Pelletier, G., Sol, H. and Asseo, E.: Magnetized Langmuir wave-packets excited by a strong beam-plasma interaction. Phys. Rev. A **38**, 2552–2563 (1988)
16. Robinson, P. A. and Newman, D. H.: Strong Langmuir turbulence generated by electron-beams – electric-field distributions and electron-scattering. Phys. Fluids B **2**, 3120–3133 (1990)
17. Simpson, G., Sulem, C. and Sulem, P. L.: Arrest of Langmuir wave collapse by quantum effects. Phys. Rev. E **80**, 056405–056414 (2009)
18. Stenflo, L.: Upper-hybrid wave collapse. Phys. Rev. Lett. **48**, 1441–1441 (1982)
19. Thornhill, S. G. and ter Haar, D.: Langmuir turbulence and modulational instability. Phys. Rep. **43**, 43–99 (1978)
20. Zakharov, V. E.: Collapse of Langmuir waves. Sov. Phys. J. Exp. Theor. Phys. **35**, 908–914 (1972)
21. Zakharov, V. E., Mastryukov, A. F. and Sinakh, V. S.: 2-dimensional collapse of Langmuir waves. J. Exp. Theor. Phys. Letters **20**, 3–4 (1974)
22. Zakharov, V. E.: Handbook of Plasma Physics, vol. 2, eds. M. N. Rosenbluth and R. Z. Sagdeev. Elsevier, New York (1984)

Chapter 9
The Moments Method

Abstract The quantum hydrodynamic model for plasmas is extended through the inclusion of higher-order moments of the Wigner function. In this manner, quantum effects appear without the need of the Madelung decomposition of the ensemble wavefunctions. The treatment apply to the electrostatic as well as the electromagnetic cases. To describe nonlinearities and inhomogeneous magnetic fields, a gauge invariant Wigner formulation is unavoidable. Dispersion relations for high-frequency wave propagation are discussed.

9.1 Moments Method

As seen in Chaps. 4 and 6, a closed set of quantum fluid equations for plasmas is obtained, supposing the sum of the kinetic and osmotic pressures to be given in terms of an equation of state. The choice of equation of state is dictated by the (quasi) equilibrium properties of the system. For instance, one can choose the equation of state corresponding to a zero-temperature Fermi–Dirac distribution. Besides, the quantum hydrodynamical model contains also quantum diffraction effects represented by the Bohm potential term.

In spite of the simplicity and physical appeal of the resulting model, the closure hypothesis and the choice of equation of state are not free of objections, as for any set of macroscopic equations. A more complete theory would include higher-order moments of the Wigner function, beyond the zeroth- and first-order moments associated with the particle and current densities, respectively. Therefore, we do not perform a Madelung decomposition to identify the different pieces in the pressure dyad (kinetic pressure + osmotic pressure + Bohm potential). Instead, in this chapter, we define the pertinent moments and compute the associated dynamics, obtaining an alternative macroscopic theory for quantum plasmas, as described in [7,8]. Later, the same approach was applied to the treatment of the ponderomotive force [16] and to the derivation of two-fluid equations [19] in spin-dependent plasmas. We start with the electrostatic case.

F. Haas, *Quantum Plasmas: An Hydrodynamic Approach*, Springer Series on Atomic, Optical, and Plasma Physics 65, DOI 10.1007/978-1-4419-8201-8_9,

9.2 Electrostatic Case

The moments approach is traditional in classical kinetic theory, since Grad's pioneering work [6]. In addition, the moments approach for quantum charged particle systems is popular in the semiconductor community [3, 20]. However, in semiconductor devices usually one has a doping profile as well as external hetero-junction potentials which makes the analysis intrinsically nonlinear, differently from plasmas where the propagation of linear waves is the first subject worth to study. Hence, it is convenient to explicitly write the moment equations in the simplest case, namely for electrostatic quantum plasmas.

Our starting point is the Wigner–Poisson system, as discussed in Chap. 2 and written here in the three-dimensional version,

$$\frac{\partial f}{\partial t} + \mathbf{v} \cdot \nabla f + \int d\mathbf{v}' \, K(\mathbf{v}' - \mathbf{v}, \mathbf{r}) f(\mathbf{v}', \mathbf{r}) = 0, \tag{9.1}$$

$$\nabla^2 \phi = \frac{e}{\varepsilon_0} \left(\int d\mathbf{v} \, f(\mathbf{v}, \mathbf{r}) - n_0 \right), \tag{9.2}$$

where $K(\mathbf{v}' - \mathbf{v}, \mathbf{r})$ is defined by

$$K(\mathbf{v}' - \mathbf{v}, \mathbf{r}) = \frac{ie}{\hbar} \left(\frac{m}{2\pi\hbar} \right)^3 \int d\mathbf{s} \exp \left[im(\mathbf{v} - \mathbf{v}') \cdot \frac{\mathbf{s}}{\hbar} \right]$$

$$\times \left[\phi \left(\mathbf{r} + \frac{\mathbf{s}}{2} \right) - \phi \left(\mathbf{r} - \frac{\mathbf{s}}{2} \right) \right]. \tag{9.3}$$

For brevity, the time-dependence of the various quantities is omitted. The symbols are all as in Chap. 2 and the three-dimensional Wigner function is

$$f(\mathbf{r}, \mathbf{v}) = N \left(\frac{m}{2\pi\hbar} \right)^3 \sum_\alpha P_\alpha \int d\mathbf{s} \exp \left(\frac{im\mathbf{v} \cdot \mathbf{s}}{\hbar} \right) \psi_\alpha^* \left(\mathbf{r} + \frac{\mathbf{s}}{2} \right) \psi_\alpha \left(\mathbf{r} - \frac{\mathbf{s}}{2} \right). \tag{9.4}$$

Unlike in Chap. 2, we now prefer to use the velocity $\mathbf{v}$ instead of the canonical momentum $\mathbf{p} = m\mathbf{v}$ because of gauge invariance issues which will become clear when treating the electromagnetic case.

To obtain macroscopic equations, let us introduce the moments

$$n = \int d\mathbf{v} f(\mathbf{r}, \mathbf{v}), \tag{9.5}$$

$$n\mathbf{u} = \int d\mathbf{v} \, \mathbf{v} \, f(\mathbf{r}, \mathbf{v}), \tag{9.6}$$

$$P_{ij} = m \left(\int d\mathbf{v} \, v_i v_j f - n u_i u_j \right), \tag{9.7}$$

$$Q_{ijk} = m \int d\mathbf{v} \, (v_i - u_i)(v_j - u_j)(v_k - u_k) f(\mathbf{r}, \mathbf{v}), \tag{9.8}$$

$$R_{ijkl} = m \int d\mathbf{v} \, (v_i - u_i)(v_j - u_j)(v_k - u_k)(v_l - u_l) f(\mathbf{r}, \mathbf{v}). \tag{9.9}$$

In special, from the pressure dyad $\mathbf{P}$ a scalar pressure P can be obtained,

$$P = \left(\frac{1}{3}\right) P_{ii}. \tag{9.10}$$

Similarly, from the triad $\mathbf{Q}$ one can form the heat flux vector $\mathbf{q}$, with

$$q_i = \left(\frac{1}{2}\right) Q_{jji}. \tag{9.11}$$

Here and in the following, the summation convention is applied.

From the moments of (9.1), one obtain the following macroscopic equations,

$$\frac{Dn}{Dt} = -n\nabla \cdot \mathbf{u}, \tag{9.12}$$

$$\frac{Du_i}{Dt} = -\frac{\partial_j P_{ij}}{mn} + \frac{e}{m}\partial_i\phi, \tag{9.13}$$

$$\frac{DP_{ij}}{Dt} = -P_{k(i}\partial^k u_{j)} - P_{ij}\nabla \cdot \mathbf{u} - \partial^k Q_{ijk}, \tag{9.14}$$

$$\frac{DQ_{ijk}}{Dt} = \frac{1}{mn} P_{(ij}\partial^l P_{k)l} - Q_{l(ij}\partial^l u_{k)} - Q_{ijk}\nabla \cdot \mathbf{u}$$

$$- \frac{e\hbar^2 n}{4m^2}\partial^3_{ijk}\phi - \partial^l R_{ijkl}, \tag{9.15}$$

where $\partial_i = \partial^i = \partial/\partial r_i$ and

$$\frac{D}{Dt} = \frac{\partial}{\partial t} + \mathbf{u} \cdot \nabla \tag{9.16}$$

is the material derivative. To close the system, we need Poisson's equation,

$$\nabla^2\phi = \frac{e}{\varepsilon_0}(n - n_0). \tag{9.17}$$

The calculation assumes, for example, decaying or periodic boundary conditions and uses that P_{ij}, Q_{ijk}, and R_{ijkl} are completely symmetric under index permutation. In addition, in (9.14) and (9.15) the round brackets denote symmetrization, where we use a minimal sum over permutations of free indices needed to get symmetric tensors. Thus, for example,

$$P_{k(i}\partial^k u_{j)} \equiv P_{ki}\partial^k u_j + P_{kj}\partial^k u_i, \tag{9.18}$$

$$P_{(ij}\partial^l P_{k)l} \equiv P_{ij}\partial^l P_{kl} + P_{jk}\partial^l P_{il} + P_{ki}\partial^l P_{jl} \tag{9.19}$$

and so on.

Presently, there is no assumption on the particular equilibrium Wigner function. This is a difference in comparison with approaches [4] relying on the first-order quantum correction to Maxwell–Boltzmann equilibria [18]. Therefore, the model is not semi-classical; there is not necessarily a small quantum parameter measuring quantum diffraction effects. Moreover, it is not restricted to classical, Maxwell–Boltzmann statistics. In addition, notice that the explicit dependence on Planck's constant appears only when the heat flux triad transport equation is considered. Finally, the quantum contribution disappears in (9.15) when the scalar potential is absent. This is similar to Gardner's approach [4] and in contrast to the Madelung-decomposition method of Chaps. 4 and 6 where quantum effects modeled by a Bohm potential appear already at the momentum transport equation. Here, the usual quantum force is replaced by the third-order derivative of the scalar potential term in (9.15). It is remarkable that in this context a free particle gas would be described by entirely classical equations.

It is to be expected from the very beginning that Planck's constant would not appear through the moments of the Wigner equation when there is no electric field because in this case the Wigner equation reduces to the free-particle Vlasov equation. However, even so Wigner functions are quantum objects, associated with a certain quantum statistical ensemble. Furthermore, Planck's constant should appear in the initial conditions for the Wigner function.

Let us examine the derivation of the moment equations. The continuity equation (9.12) and the force equation (9.13) follows in a straightforward manner taking the zeroth- and first-order moments of (9.3) for the Wigner function. To obtain (9.14) for the pressure dyad, multiply each term of (9.3) by $v_i v_j$ and integrate to get

$$\frac{\partial}{\partial t} \int d\mathbf{v}\, v_i v_j f(\mathbf{r}, \mathbf{v}) + \partial_k \int d\mathbf{v}\, v_i v_j v_k f(\mathbf{r}, \mathbf{v}) + \int d\mathbf{v}' K(\mathbf{v}' - \mathbf{v}) v_i v_j f(\mathbf{r}, \mathbf{v}) = 0.$$
(9.20)

The first two integrals above can be handled considering

$$\int d\mathbf{v}\, v_i v_j f(\mathbf{r}, \mathbf{v}) = \frac{P_{ij}}{m} + n u_i u_j,$$
(9.21)

$$\int d\mathbf{v}\, v_i v_j v_k f(\mathbf{r}, \mathbf{v}) = \frac{Q_{ijk}}{m} + \frac{u_{(i} P_{jk)}}{m} + n u_i u_j u_k,$$
(9.22)

which can be proven by expanding P_{ij} and Q_{ijk} in terms of the lower-order moments. The third term in (9.20) is

$$\int d\mathbf{v}' K(\mathbf{v}' - \mathbf{v}) v_i v_j f(\mathbf{r}, \mathbf{v}) = -\frac{e n}{m} u_{(i} \partial_{j)} \phi,$$
(9.23)

as follows from the identity

$$\left(\frac{m}{2\pi\hbar}\right)^3 \int d\mathbf{v} \exp\left(\frac{i m \mathbf{v} \cdot \mathbf{s}}{\hbar}\right) = -\frac{\hbar^2}{m^2} \frac{\partial^2}{\partial s_i \partial s_j} \delta(\mathbf{s})$$
(9.24)

together with integration by parts. Joining the results and using the continuity and force equations to eliminate $\partial n/\partial t$ and $\partial \mathbf{u}/\partial t$ after some algebra, we arrive at (9.14). Similarly, the transport equation (9.15) for the heat triad Q_{ijk} is found.

9.3 Dispersion Relation for Electrostatic Waves

We disregard the contribution from the fourth-order moment R_{ijkl}, which is the simplest way to achieve closure of the system (9.12)–(9.15). We take into account Poisson's equation and linearize around the homogeneous equilibrium $n = n_0$, $\mathbf{u} = 0$,

$$P_{ij} = n_0 \kappa_B \left[T_{0\perp} (\hat{\mathbf{x}} \otimes \hat{\mathbf{x}} + \hat{\mathbf{y}} \otimes \hat{\mathbf{y}}) + T_{0\parallel} \hat{\mathbf{z}} \otimes \hat{\mathbf{z}} \right], \tag{9.25}$$

$Q_{ijk} = 0$, and $\phi = 0$, where the equilibrium temperatures perpendicular and parallel to the wave propagation $T_{0\perp}$ and $T_{0\parallel}$ can be unequal. Here, κ_B is Boltzmann's constant and plane wave perturbations proportional to $\exp(ikz - i\omega t)$ are assumed, without loss of generality. The wave propagating in the z-axis direction is an anisotropy source. It follows that

$$\omega^2 = \frac{\omega_p^2}{2} \left[1 + \left(1 + \frac{12 \kappa_B T_{0\parallel} k^2}{m \omega_p^2} + \frac{\hbar^2 k^4}{m^2 \omega_p^2} \right)^{1/2} \right], \tag{9.26}$$

where $\omega_p = (n_0 e^2 / m \varepsilon_0)^{1/2}$ is the plasma frequency. The transverse temperature $T_{0\perp}$ does not contribute. To obtain (9.26), we need to consider the linearization of the moment equations, which yield an homogeneous system of 20 equations for 20 unknowns composing the perturbations δn, δu_i, δP_{ij}, and δQ_{ijk}, taking into account the symmetrization.

In the particular case of small wavenumber and quantum effects, (9.26) reduces to the usual quantum Langmuir dispersion relation

$$\omega^2 = \omega_p^2 + \frac{3 \kappa_B T_{0\parallel} k^2}{m} + \frac{\hbar^2 k^4}{4 m^2}. \tag{9.27}$$

The temperature $T_{0\parallel}$ should be associated with the velocities dispersion of the equilibrium Wigner function. For Maxwell–Boltzmann equilibrium, one has $T_{0\parallel} = T$, the thermodynamic temperature. For a zero-temperature Fermi gas, one has $T_{0\parallel} = (2/5)T_F$, where T_F is the Fermi temperature.

Contrarily to the habitual prejudice, one cannot simply postulate $P_{ij} = P \delta_{ij}$ in terms of the scalar pressure P because the linearization of (9.13)–(9.15) with $P_{ij} = P_{0ij} + \delta P_{ij}$, $\phi = \delta \phi$ gives

$$\delta P_{ij} = -\frac{e \, \delta \phi \, k^2}{m \omega^2} \left(P_{0ij} + P_{0(iz} \delta_{jz)} + \frac{n_0 \hbar^2 k^2 \delta_{iz} \delta_{jz}}{4 m} \right) \tag{9.28}$$

as the first-order perturbation of the pressure dyad, assuming $\mathbf{k} = k\hat{z}$ and where the Kronecker delta δ_{ij} was employed. Clearly, the wave propagation itself is a source of anisotropy because there will be nonzero off-diagonal components of the perturbed pressure dyad even for initially isotropic equilibria. The result (9.28) also follows from the kinetic theory in the long wavelength limit, and is independent of the classical or quantum nature of the system. However, as apparent from (9.28), it is acceptable to assume

$$P_{ij} = n\kappa_{B}\left[T_{\perp}\left(\hat{\mathbf{x}}\otimes\hat{\mathbf{x}}+\hat{\mathbf{y}}\otimes\hat{\mathbf{y}}\right)+T_{\parallel}\hat{\mathbf{z}}\otimes\hat{\mathbf{z}}\right], \tag{9.29}$$

where the transverse and parallel temperatures $T_{\perp}$ and $T_{\parallel}$ in general are different. Anisotropic forms of the pressure dyad are usually considered in the case of strong magnetic fields; here, we show that they should be followed also in the electrostatic case because of the wave propagation which in itself defines a preferred direction.

The Bohm–Gross dispersion relation is recovered from (9.12)–(9.14) in the adiabatic and classical case. When heat transfer is irrelevant, we can postulate $Q_{ijk} = 0$ and forget (9.15). Linearizing the remaining equations, the result is

$$\omega^2 = \omega_p^2 + 3\left(\frac{\kappa_B T_{0\parallel}}{m}\right)k^2, \tag{9.30}$$

where only the parallel component $P_{0zz} \equiv n_0\,\kappa_B\,T_{0\parallel}$ of the equilibrium pressure dyad contributes. However, if the transport equation for the third-order moment Q_{ijk} is not included no quantum effects are present!

Insisting on an isotropic pressure dyad $P_{ij} = P\,\delta_{ij}$ and taking $Q_{ijk} = 0$, one obtain, in particular,

$$\frac{DP}{Dt} = -\frac{5}{3}P\nabla\cdot\mathbf{u}, \tag{9.31}$$

as follows after taking the trace of all terms in (9.14). Using the continuity equation one then derive

$$\frac{D}{Dt}\left(Pn^{-5/3}\right) = 0, \tag{9.32}$$

which is consistent with the classical, adiabatic equation of state $P \sim n^{5/3}$ as expected.

Alternative moment hierarchy formulations [20], closed at the temperature (basically the trace of the second-order moment of the Wigner function) evolution equation, can be shown to result in $\omega^2 = \omega_p^2 + (5/3)\,(\kappa_B T_0/m)\,k^2 + \hbar^2 k^4/(12\,m^2)$, which goes neither to the Bohm–Gross nor Bohm–Pines dispersion relations in the classical or zero-temperature limits, respectively. In contrast, as shown here, the quantum modified Bohm-Gross dispersion relation is a natural consequence from third-order moment theory, in the small wavenumber and quantum effects limit of (9.26).

In addition to the above linear wave propagation analysis, nonlinear oscillating solutions for the electrostatic moment hierarchy equations (9.12)–(9.15) and (9.17) can be found in [7]. Next, we investigate the electromagnetic case.

9.3.1 Electromagnetic Case

Frequently Wigner function methods are applied only to electrostatic problems. One reason for this is the considerable complexity of the quantum Vlasov equation including magnetic fields. Already in the electrostatic case, the quantum Vlasov equation include a cumbersome integro-differential term, so that hardly it can be examined except in the linear limit or numerically. However, the emergence of new areas like spintronics [21] where magnetic effects are crucial makes it desirable to have quantum kinetic models allowing for nonzero vector potentials, besides the intrinsic interest of the subject. In addition, a gauge invariant formalism is needed to avoid inconsistencies, as will be shortly verified.

The gauge invariant Wigner function (GIWF) formalism have already been detailed in the past [11, 13, 17]. Reference [13] by Serimaa et al. contain an elegant expression for the evolution equation satisfied by the GIWF, see (9.39). The gauge-free Wigner function method has been applied in describing friction as a result of radiation reaction [9]. Also notice the work by Bialynicki-Birula et al. [1], where the quantum phase-space equations have been applied and explicitly solved for spinning particles in a gauge invariant manner. Here we use the GIWF method to model quantum plasmas, as proposed for the first time in [8].

Since the electromagnetic quantum Vlasov equation is quite complicated, we introduce a macroscopic theory taking the moments of the GIWF, as done in the electrostatic case. As stressed in this monograph, in macroscopic models the nonlinear regimes are not necessarily unaccessible to qualitative analysis. Also, they are less numerically demanding. The price of replacing the more detailed kinetic models by macroscopic models is the loss of information on kinetic phenomena like Landau damping, the plasma echo, particle trapping, etc.

9.4 Gauge Invariant Wigner Function

A sensible definition of gauge invariant one-particle Wigner function $f = f(\mathbf{r}, \mathbf{v})$ was introduced by Stratonovich [17]. In a noncovariant form it is given by

$$
f(\mathbf{r}, \mathbf{v}) = N \left(\frac{m}{2\pi\hbar} \right)^3 \int d\mathbf{s} \exp \left[\frac{i\mathbf{s}}{\hbar} \cdot \left(m\mathbf{v} - e \int_{-1/2}^{1/2} d\tau \mathbf{A}(\mathbf{r} + \tau\mathbf{s}) \right) \right]
$$
$$
\times \psi^* \left(\mathbf{r} + \frac{\mathbf{s}}{2} \right) \psi \left(\mathbf{r} - \frac{\mathbf{s}}{2} \right), \tag{9.33}
$$

to be compared with the definition (9.4) in the electrostatic case. For brevity, the time-dependence of all quantities is omitted. To simplify the notation, a pure state, normalized to unity wavefunction is used, although the results hold equally well in the mixed state case. In addition, $\mathbf{A}(\mathbf{r})$ is the vector potential. The charge of the plasma particles is $-e$.

It is immediate to verify that the extra integral in (9.33) containing the vector potential compensates for the change in the wavefunction in a local gauge transformation

$$\mathbf{A} \to \mathbf{A} + \nabla \Lambda, \quad \psi \to \psi \exp\left(\frac{-ie\Lambda}{\hbar}\right), \tag{9.34}$$

where $\Lambda = \Lambda(\mathbf{r})$ is an arbitrary differentiable function. Indeed,

$$\mathbf{s} \cdot \int_{-1/2}^{1/2} d\tau \nabla \Lambda(\mathbf{r} + \tau \mathbf{s}) = \Lambda\left(\mathbf{r} + \frac{\mathbf{s}}{2}\right) - \Lambda\left(\mathbf{r} - \frac{\mathbf{s}}{2}\right). \tag{9.35}$$

Our choice of the form (9.33) is due to convenience since it provides a nonambiguous way to calculate averaged quantities. The phase factor in it can be justified [11] in terms of the minimal coupling principle. Moreover, as discussed in more detail elsewhere, the function of the phase factor is to convert any gauge into the axial gauge [11]. However, there are other ways to introduce gauge-free Wigner functions. For instance, one can take [2] a GIWF written in terms of a line integral $\int_{\mathbf{r}_1}^{\mathbf{r}_2} \mathbf{A}(\mathbf{s}, t) \cdot d\mathbf{s}$ instead of the chosen phase factor. However, in this case how to chose the integration path from $\mathbf{r}_1$ to $\mathbf{r}_2$? To avoid ambiguities, we apply (9.33).

From f, we can compute the very basic zeroth- and first-order moments

$$\int d\mathbf{v} f = N|\psi|^2, \tag{9.36}$$

$$\int d\mathbf{v}\, \mathbf{v} f = \frac{i\hbar N}{2m}(\psi \nabla \psi^* - \psi^* \nabla \psi) + \frac{Ne}{m}|\psi|^2 \mathbf{A}, \tag{9.37}$$

with the interpretation of particle and current densities, respectively. By construction these quantities are invariant under local gauge transformations. The same apply if the usual (gauge dependent) Wigner function

$$f^{GD}(\mathbf{r}, \mathbf{p}) = \frac{N}{(2\pi\hbar)^3} \int d\mathbf{s} \exp\left(\frac{i\mathbf{p}\cdot\mathbf{s}}{\hbar}\right) \psi^*\left(\mathbf{r} + \frac{\mathbf{s}}{2}\right) \psi\left(\mathbf{r} - \frac{\mathbf{s}}{2}\right), \tag{9.38}$$

is employed, where the canonical moment is $\mathbf{p} = m\mathbf{v} + e\mathbf{A}$. Also the second-order moments from f^{GD} are gauge-free, the troubles starting when considering higher-order moments. Serious drawbacks occurs when calculating the evolution equation for the second-order moment of the usual Wigner function, as will be seen later.

The time-evolution of the GIWF was considered already by Stratonovich [17], but a particularly illuminating form to express it was provided by Serimaa et al. [13]:

$$\left\{ \frac{\partial}{\partial t} + (\mathbf{v} + \Delta \tilde{\mathbf{v}}) \cdot \frac{\partial}{\partial \mathbf{r}} - \frac{e}{m} \left[\tilde{\mathbf{E}} + (\mathbf{v} + \Delta \tilde{\mathbf{v}}) \times \tilde{\mathbf{B}} \right] \cdot \frac{\partial}{\partial \mathbf{v}} \right\} f(\mathbf{r}, \mathbf{v}) = 0. \qquad (9.39)$$

Here, we introduced the operators

$$\Delta \tilde{\mathbf{v}} = -\frac{i\hbar e}{m^2} \frac{\partial}{\partial \mathbf{v}} \times \int_{-1/2}^{1/2} d\tau\, \tau \mathbf{B} \left(\mathbf{r} + \frac{i\hbar \tau}{m} \frac{\partial}{\partial \mathbf{v}} \right), \qquad (9.40)$$

$$\tilde{\mathbf{E}} = \int_{-1/2}^{1/2} d\tau\, \mathbf{E} \left(\mathbf{r} + \frac{i\hbar \tau}{m} \frac{\partial}{\partial \mathbf{v}} \right), \qquad (9.41)$$

$$\tilde{\mathbf{B}} = \int_{-1/2}^{1/2} d\tau\, \mathbf{B} \left(\mathbf{r} + \frac{i\hbar \tau}{m} \frac{\partial}{\partial \mathbf{v}} \right), \qquad (9.42)$$

where $\mathbf{B} = \mathbf{B}(\mathbf{r})$ and $\mathbf{E} = \mathbf{E}(\mathbf{r})$ are the magnetic and electric fields, respectively. The kinetic equation (9.39) follows from the Schrödinger equation for the wavefunction or, alternatively, from the von Neumann equation solved by the density matrix [13]. Hence, we have a manifestly gauge-free formalism, where the time-evolution of the Wigner function involves only the physical fields.

The operators in (9.40)–(9.42) are understood in terms of the series expansion of the integrands in powers of $\hbar$ (so that the electric and magnetic fields are supposed to be differentiable). After Taylor expanding, one perform the τ integration.

Equation (9.39) resembles Vlasov's equation, with two differences: the electromagnetic fields are replaced by $\tilde{\mathbf{E}}$ and $\tilde{\mathbf{B}}$ defined in (9.41) and (9.42); the velocity vector is displaced by the intrinsically quantum mechanical perturbation $\Delta \tilde{\mathbf{v}}$ defined in (9.40). This perturbation $\Delta \tilde{\mathbf{v}}$ is zero in the electrostatic case. Moreover, not any function can be taken as a Wigner function, see conditions (2.56)–(2.59).

9.5 Macroscopic Equations

Equation (9.39) is in a compact form, but it becomes quite complicated after explicitly writing the operators $\Delta \tilde{\mathbf{v}}$, $\tilde{\mathbf{E}}$ and $\tilde{\mathbf{B}}$. Hence, it is useful to consider a moments formulation. We define the moments as in (9.5)–(9.9), where now f is the GIWF.

For the sake of calculating the moments hierarchy equations, it is convenient to expand $\Delta \tilde{\mathbf{v}}$, $\tilde{\mathbf{B}}$ and $\tilde{\mathbf{E}}$ according to

$$\Delta \tilde{v}_i = -\frac{q\hbar^2 \varepsilon_{ijk}}{12m^3} \partial_m B_k \frac{\partial^2}{\partial v_j \partial v_m} + \frac{q\hbar^4 \varepsilon_{ijk}}{540m^5} \partial_{mnl}^3 B_k \frac{\partial^4}{\partial v_j \partial v_m \partial v_n \partial v_l} + \cdots, \qquad (9.43)$$

$$\tilde{E}_i = E_i - \frac{\hbar^2}{24m^2} \partial_{jk}^2 E_i \frac{\partial^2}{\partial v_j \partial v_k} + \frac{\hbar^4}{1920m^4} \partial_{jkmn}^4 E_i \frac{\partial^4}{\partial v_j \partial v_k \partial v_m \partial v_n} + \cdots, \qquad (9.44)$$

$$\tilde{B}_i = B_i - \frac{\hbar^2}{24m^2} \partial_{jk}^2 B_i \frac{\partial^2}{\partial v_j \partial v_k} + \frac{\hbar^4}{1920m^4} \partial_{jkmn}^4 B_i \frac{\partial^4}{\partial v_j \partial v_k \partial v_m \partial v_n} + \cdots. \qquad (9.45)$$

disregarding higher-order quantum corrections. The notation $\partial_i \equiv \partial/\partial r_i$ is used. Equations (9.43)–(9.45) are proven directly from the definitions of $\Delta\tilde{\mathbf{v}}$, $\tilde{\mathbf{E}}$, and $\tilde{\mathbf{B}}$.

Assuming decaying boundary conditions, as far as the moment hierarchy is closed at the third-rank stress tensor, only the leading quantum corrections [the terms $\propto \hbar^2$ in (9.43)–(9.45)] are needed. This is due to the structure of the higher-order corrections. Indeed, these terms always involve at least fourth-order velocity derivatives and, for instance,

$$\int d\mathbf{v}\, v_i v_j v_k \frac{\partial^4 f}{\partial v_a \partial v_b \partial v_c \partial_d} = 0. \tag{9.46}$$

Therefore, only the semiclassical Wigner equation is needed, which does not mean that the quantum effects are necessarily small. It just happens that higher-order quantum corrections would appear only for higher-order moment evolution equations.

Following (9.39), the semiclassical electromagnetic Wigner equation then reads

$$\left[\frac{\partial}{\partial t} + \mathbf{v} \cdot \frac{\partial}{\partial \mathbf{r}} - \frac{e}{m}\left(\mathbf{E} + \mathbf{v} \times \mathbf{B}\right) \cdot \frac{\partial}{\partial \mathbf{v}} \right] f(\mathbf{r},\mathbf{v}) = -\frac{e\hbar^2}{24m^3} \partial_{jk}^2 E_i \frac{\partial^3 f}{\partial v_i \partial v_j \partial v_k}$$

$$-\frac{e\hbar^2 \varepsilon_{ijk}}{12m^3} \partial_m B_k \frac{\partial^3 f}{\partial r_i \partial v_j \partial v_m} - \frac{e\hbar^2 \varepsilon_{ijk} v_j}{24m^3} \partial_{mn}^2 B_k \frac{\partial^3 f}{\partial v_i \partial v_m \partial v_n}$$

$$+\frac{e^2\hbar^2}{12m^4}\left(B_i \partial_j B_k \frac{\partial^3 f}{\partial v_i \partial v_j \partial v_k} - B_i \partial_j B_i \frac{\partial^3 f}{\partial v_j \partial v_k \partial v_k} \right). \tag{9.47}$$

After a tedious algebra, the result for the moments hierarchy equation is

$$\frac{Dn}{Dt} + n\nabla \cdot \mathbf{u} = 0, \tag{9.48}$$

$$\frac{Du_i}{Dt} = -\frac{\partial_j P_{ij}}{mn} - \frac{e}{m}\left(\mathbf{E} + \mathbf{u} \times \mathbf{B}\right)_i, \tag{9.49}$$

$$\frac{DP_{ij}}{Dt} = -P_{ik}\,\partial_k u_j - P_{jk}\,\partial_k u_i - P_{ij}\nabla \cdot \mathbf{u} - \frac{e}{m}\varepsilon_{imn}P_{jm}B_n - \frac{e}{m}\varepsilon_{jmn}P_{im}B_n$$

$$-\frac{e\hbar^2}{12m^2}\varepsilon_{ikl}\partial_l\left(n\partial_j B_k\right) - \frac{e\hbar^2}{12m^2}\varepsilon_{jkl}\partial_l\left(n\partial_i B_k\right) - \partial_k Q_{ijk}, \tag{9.50}$$

$$\frac{DQ_{ijk}}{Dt} = -Q_{ijr}\,\partial_r u_k - Q_{jkr}\,\partial_r u_i - Q_{kir}\,\partial_r u_j - Q_{ijk}\nabla \cdot \mathbf{u} - \partial_r R_{ijkr}$$

$$+\frac{1}{mn}\left(P_{ij}\partial_r P_{kr} + P_{jk}\partial_r P_{ir} + P_{ki}\partial_r P_{jr} \right) - \frac{e}{m}\left(\varepsilon_{irs}Q_{rjk} + \varepsilon_{jrs}Q_{rki} + \varepsilon_{krs}Q_{rij} \right)B_s$$

$$+\frac{e\hbar^2 n}{12m^2}\left(\partial_{ij}^2 E_k + \partial_{jk}^2 E_i + \partial_{ki}^2 E_j \right) + \frac{e^2\hbar^2 n}{12m^3}\left(\delta_{ij}\partial_k + \delta_{jk}\partial_i + \delta_{ki}\partial_j \right)B^2$$

$$+\frac{e\hbar^2 n}{12m^2}\left[(\mathbf{u}\times\partial_{jk}^2\mathbf{B})_i+(\mathbf{u}\times\partial_{ki}^2\mathbf{B})_j+(\mathbf{u}\times\partial_{ij}^2\mathbf{B})_k\right]$$

$$-\frac{e\hbar^2 n}{12m^2}\left[\varepsilon_{irs}\left(\partial_j B_r\partial_s u_k+\partial_k B_r\partial_s u_j\right)+\varepsilon_{jrs}\left(\partial_k B_r\partial_s u_i+\partial_i B_r\partial_s u_k\right)\right.$$

$$\left.+\varepsilon_{krs}\left(\partial_i B_r\partial_s u_j+\partial_j B_r\partial_s u_i\right)\right]$$

$$-\frac{e^2\hbar^2 n}{12m^3}\left[\partial_i(B_j B_k)+\partial_j(B_k B_i)+\partial_k(B_i B_j)\right]. \tag{9.51}$$

When $\mathbf{B}=0$, (9.48)–(9.51) recover the electrostatic equations (9.12)–(9.15). In the formal classical limit $\hbar\to 0$, the classical electromagnetic moment hierarchy equations are recovered [5, 12, 15]. Now quantum effects are explicit already in the transport equation for the pressure dyad, through the magnetic field.

If we have employed the traditional, gauge dependent Wigner function, it would not be possible to proceed exactly as in the classical case in the definition of the moments. Indeed, it would then necessary to define them as

$$n=\int d\mathbf{p} f^{GD}, \tag{9.52}$$

$$n\mathbf{u}=\int d\mathbf{p}\left(\frac{\mathbf{p}+e\mathbf{A}}{m}\right)f^{GD}, \tag{9.53}$$

$$P_{ij}=m\left(\int d\mathbf{p}\frac{(p_i+eA_i)(p_j+eA_j)}{m^2}f^{GD}-n u_i u_j\right), \tag{9.54}$$

$$Q_{ijk}^{GV}=\frac{1}{m^2}\int d\mathbf{p}\,(p_i+eA_i-mu_i)(p_j+eA_j-mu_j)(p_k+eA_k-mu_k)f^{GV}. \tag{9.55}$$

The same symbols n, $\mathbf{u}$ and P_{ij} are used on purpose since (9.52)–(9.54) produce the same expressions as from the GIWF, in spite of the fact that f^{GV} itself is a gauge dependent object. However, from (6.3) satisfied by the usual Wigner equation one would obtain

$$\frac{DP_{ij}}{Dt}=-P_{ik}\partial_k u_j-P_{jk}\partial_k u_i-P_{ij}\nabla\cdot\mathbf{u}-\frac{e}{m}\varepsilon_{imn}P_{jm}B_n-\frac{e}{m}\varepsilon_{jmn}P_{im}B_n$$

$$+\frac{e\hbar^2}{4m^2}\partial_{ij}^2\mathbf{A}\cdot\nabla n-\partial_k Q_{ijk}^{GV}, \tag{9.56}$$

which is not gauge-free. The reason is that

$$Q_{ijk}^{GV}=Q_{ijk}+\frac{e\hbar^2 n}{12m^2}\left(\partial_{ij}^2 A_k+\partial_{jk}^2 A_i+\partial_{ki}^2 A_j\right) \tag{9.57}$$

is not gauge invariant. If Q_{ijk}^{GV} from (9.57) is inserted into (9.56) one rederive (9.50) for the pressure dyad, but only assuming the Coulomb gauge $\nabla\cdot\mathbf{A}=0$. Also transport equations for the higher-order moments are not gauge invariant. For this reason, we do not use the usual gauge-dependent Wigner function anymore.

9.6   Electromagnetic Dispersion Relation

The fluid model given by (9.48)–(9.51) can be used to describe linear transverse waves. Considering an one-component plasma, with an homogeneous neutralizing ionic background with density n_0. The system is then closed by Maxwell equations,

$$\nabla \cdot \mathbf{E} = \frac{e}{\varepsilon_0}(n - n_0), \quad \nabla \cdot \mathbf{B} = 0, \tag{9.58}$$

$$\nabla \times \mathbf{E} = -\frac{\partial \mathbf{B}}{\partial t}, \quad \nabla \times \mathbf{B} = -\mu_0 e n \mathbf{u} + \frac{1}{c^2}\frac{\partial \mathbf{E}}{\partial t}. \tag{9.59}$$

The moment equations can be linearized around the equilibrium

$$n = n_0, \mathbf{u} = 0, P_{ij} = P_{ij}^{(0)}, Q_{ijk} = 0, R_{ijkl} = 0, \mathbf{E} = 0, \mathbf{B} = 0. \tag{9.60}$$

To consider waves propagating in the z-direction with transverse polarization we let all fluctuations have the space-time dependence $e^{ikz - i\omega t}$ and set $E_z = 0$. Moreover, we decompose the zeroth-order pressure dyad as

$$P_{ij}^{(0)} = P_\perp(\delta_{ix}\delta_{jx} + \delta_{iy}\delta_{jy}) + P_\parallel \delta_{iz}\delta_{jz}, \tag{9.61}$$

allowing for anisotropy, where $P_\perp$ and $P_\parallel$ are constants.

It turns out that if we use the closure assumption $R_{ijkl} = 0$ the quantum corrections to the transverse modes will not be retained so that to display the lowest-order quantum corrections, it is necessary to take into account also the contribution from the fourth-order moment. As a closure assumption, we use

$$R_{ijkl} = \frac{e\hbar^2}{4m^3\omega^2}\left(P_{im}^{(0)}\partial_{jkl}^3 + P_{jm}^{(0)}\partial_{kli}^3 + P_{km}^{(0)}\partial_{lij}^3 + P_{lm}^{(0)}\partial_{ijk}^3\right)E_m, \tag{9.62}$$

adapted to the transverse wave case. The closure (9.62) is deduced systematically from the linearized equations satisfied by the fourth- and fifth-order moments, see [8]. Note that in principle the fourth-order moment R_{ijkl} can have a nonzero equilibrium contribution $R_{ijkl}^{(0)} \sim v_T^4$, where $v_T = \sqrt{(2P_\perp + P_\parallel)/(mn_0)}$ is the thermal velocity, but we will neglect this since we are looking only for the lowest-order correction. Likewise for the terms $\sim \hbar^4$. Finally, it is worth to remark that in the classical limit the fourth-order moment could be set to zero.

The linearized equations can then be solved by first writing the magnetic field in terms of the electric field and then eliminating all quantities except the velocity so that we obtain the velocity in terms of the electric field. Coupling the resulting equation with Faraday's law the dispersion relation

$$\omega^2 - k^2 c^2 = \omega_p^2\left[1 + \frac{k^2 P_\perp}{n_0 m \omega^2} + \frac{\hbar^2 k^6 P_\perp}{4n_0 m^3 \omega^4}\right], \tag{9.63}$$

is obtained. Here, $\omega_p = \sqrt{n_0 e^2/(m\varepsilon_0)}$ is the plasma frequency. If, instead, the closure $R_{ijkl} = 0$ was used, the term proportional to $\hbar^2$ would be absent in the dispersion relation.

In the simultaneous long wavelength and semiclassical limits, (9.63) can be shown to admit an approximate solution

$$\omega^2 \simeq \omega_p^2 + c^2 k^2 + \frac{P_\perp k^2}{m n_0} + \frac{\hbar^2 k^6 P_\perp}{4 m^3 n_0 \omega_p^2}. \tag{9.64}$$

To check the consistency, we need to compare to the results from kinetic theory. Here, we are not concerned with Landau damping issues so that all integrals can be interpreted in the principal value sense. Assume

$$\mathbf{E} = \mathbf{E}_1 \exp[i(kz - \omega t)], \tag{9.65}$$

$$\mathbf{B} = \mathbf{B}_1 \exp[i(kz - \omega t)], \tag{9.66}$$

$$f = f_0(\mathbf{v}) + f_1(\mathbf{v}) \exp[i(kz - \omega t)], \tag{9.67}$$

where $\mathbf{k} \cdot \mathbf{E} = 0$ as before and with the subscript 1 denoting first-order quantities. The equilibrium Wigner function satisfy

$$\int d\mathbf{v}\, f_0 = n_0, \quad \int d\mathbf{v}\, \mathbf{v}\, f_0 = 0. \tag{9.68}$$

Further, we assume an equilibrium Wigner function such that $f_0 = f_0(v_\perp, v_z)$, where $v_\perp^2 = v_x^2 + v_y^2$. Notice that since there is no zeroth-order magnetic field the perturbation velocity $\Delta\tilde{\mathbf{v}}$ is also of first-order. Hence, $\Delta\tilde{\mathbf{v}}$ does not contribute in the linearized Wigner equation (9.39). Using (9.41) and (9.42) we get

$$\tilde{\mathbf{E}} = \mathbf{E}L, \quad \tilde{\mathbf{B}} = \mathbf{B}L, \tag{9.69}$$

defining the operator

$$L = \frac{\sinh\theta}{\theta}, \quad \theta = \frac{\hbar k}{2m}\frac{\partial}{\partial v_z}. \tag{9.70}$$

We note that

$$L\left(\frac{\partial f_0}{\partial v_z}\right) = \frac{m}{\hbar k}\left[f_0\left(\mathbf{v} + \frac{\hbar \mathbf{k}}{2m}\right) - f_0\left(\mathbf{v} - \frac{\hbar \mathbf{k}}{2m}\right)\right], \tag{9.71}$$

where $\mathbf{k} = k\hat{\mathbf{z}}$. Moreover $L \to 1$ in the classical limit, since

$$L = \sum_{j=0}^{\infty} \frac{1}{(2j+1)!}\left(\frac{\hbar k}{2m}\frac{\partial}{\partial v_z}\right)^{2j} = 1 + \frac{1}{24}\left(\frac{\hbar k}{m}\right)^2\frac{\partial^2}{\partial v_z^2} + \cdots \tag{9.72}$$

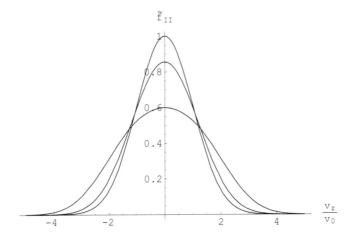

Fig. 9.1 Quantum diffusion on the equilibrium Wigner function $f_0 = f_T(v_\perp) \exp[-v_z^2/(2v_0^2)]$, according to [8]. Here $\tilde{f}_\parallel = L(\exp[-v_z^2/(2v_0^2)])$. Values of the parameter $H = \hbar k/(2mv_0)$ are $H = 0, 1$ and 2, so that $\tilde{f}_\parallel(0) = 1, 0.86$ and 0.60, respectively

Linearizing the quantum Vlasov equation (9.39) and the Maxwell equations, the result is

$$\omega^2 = \omega_p^2 + c^2 k^2 + \frac{k^2 \omega_p^2}{2n_0} \int d\mathbf{v} \, \frac{v_\perp^2 L f_0}{(\omega - \mathbf{k} \cdot \mathbf{v})^2}, \tag{9.73}$$

where c is the speed of light and ω_p is the plasma frequency. In comparison to the classical transverse dispersion relation, the only change is the replacement $f_0 \to \tilde{f}_0 = L f_0$. In a classical picture, it is as if the particle velocities were reorganized through the diffusive operator L. Also notice that still $\tilde{f}_0 = \tilde{f}_0(v_\perp, v_z)$. Moreover, the quantum diffusion induced by the operator L preserves the number of particles, since $\int d\mathbf{v} \tilde{f}_0 = \int d\mathbf{v} f_0$ due to (9.72) under decaying boundary conditions. Figure 9.1 shows the effect of L on the equilibrium $f_0 = f_T(v_\perp) \exp[-v_z^2/(2v_0^2)]$, for different values of the nondimensional parameter $H = \hbar k/(2mv_0)$. In the simultaneous long wavelength and semiclassical limits and retaining only the leading $\sim v_T^2$ thermal corrections, (9.64) and (9.73) give the same result via the natural identification $P_\perp = (m/2) \int d\mathbf{v} v_\perp^2 f_0$. This conclude the equivalence between the moments and kinetic theories, in the fluid limit. It can also be shown that the usual Wigner equation gives the same linear dispersion relation for transverse waves as the GIWF [8, 10, 14], at least for homogeneous equilibria. Hence, the gauge-invariance issues tend to be crucial only in nonlinear regimes. However, in the case of nonhomogeneous equilibria the use of a gauge independent electromagnetic Wigner equation is advisable even for linear waves.

Problems

9.1. Demonstrate (9.15) for the transport of the third-order moment Q_{ijk} in the electrostatic case.

9.2. Demonstrate (9.26), the dispersion relation for linear electrostatic waves in the moments theory.

9.3. Check (9.28) for the perturbed pressure dyad.

9.4. Linearize the electrostatic fluid moment hierarchy equations in the adiabatic ($Q_{ijk} = 0$) and classical case. Recover the Bohm–Gross dispersion relation.

9.5. Verify that the Wigner function in (9.33) is indeed gauge-free.

9.6. Prove the series expansions in (9.43)–(9.45).

9.7. Obtain the semiclassical equation (9.47) satisfied by the GIWF.

9.8. Derive the moment equations (9.48)–(9.51) valid in the electromagnetic case.

References

1. Bialynicki-Birula, I., Górnicki, P. and Rafelski, J.: Phase-space structure of the Dirac vacuum. Phys. Rev. D **44**, 1825–1836 (1991)
2. Carruthers, P. and Zachariasen, F.: Quantum collision theory with phase-space distributions. Rev. Mod. Phys. **55**, 245–285 (1983)
3. Degond, P. and Ringhofer, C.: Quantum moment hydrodynamics and the entropy principle, J. Stat. Phys. **112**, 587–628 (2003)
4. Gardner, C.L.: The quantum hydrodynamic model for semiconductor devices. SIAM J. Appl. Math. **54**, 409–427 (1994)
5. Goswami, P., Passot, T. and Sulem, P. L.: A Landau fluid model for warm collisionless plasmas. Phys. Plasmas **12**, 102109–102118 (2005)
6. Grad, H.: On the kinetic theory of rarefied gases. Commun. Pure Appl. Math. **2**, 331–407 (1949)
7. Haas, F., Marklund, M., Brodin, G. and Zamanian, J.: Fluid moment hierarchy equations derived from quantum kinetic theory. Phys. Lett. A **374**, 481–484 (2010)
8. Haas, F., Zamanian, J., Marklund, M. and Brodin, G.: Fluid moment hierarchy equations derived from gauge invariant quantum kinetic theory. New J. Phys. **12**, 073027–073040 (2010)
9. Javanainen, J., Varró, S. and Serimaa, O. T.: Gauge-independent Wigner functions. II. Inclusion of radiation reaction Phys. Rev. A **35**, 2791–2805 (1987)
10. Kuzelev, M. V. and Rukhadze, A. A.: On the quantum description of the linear kinetics of a collisionless plasma. Phys. Usp. **42**, 603–605 (1999)
11. Levanda, M. and Fleurov, V.: A Wigner quasi-distribution function for charged particles in classical electromagnetic fields. Annals of Phys. **292**, 199–231 (2001)
12. Ramos, J. J.: Fluid formalism for collisionless magnetized plasmas. Phys. Plasmas **12**, 052102–052116 (2005)
13. Serimaa, O. T., Javanainen, J. and Varró, S.: Gauge-independent Wigner functions: General formulation. Phys. Rev. A **33**, 2913–2927 (1986)

14. Silin, V. P. and Rukhadze, A. A.: Elektromagnitnye Svoystva Plazmy i Plazmopodobnykh Sred. Moscow, Gosatomizdat (1961)
15. Siregar, E. and Goldstein, M. L.: A Vlasov moment description of cyclotron waveparticle interactions. Phys. Plasmas **3**, 1437–1447 (1996)
16. Stefan, M., Zamanian, J., Brodin, G., Misra, A. P. and Marklund, M.: Ponderomotive force due to the intrinsic spin in extended fluid and kinetic models. Phys. Rev. E **83**, 036410–036416 (2011)
17. Stratonovich, R. L.: A gauge invariant analog of the Wigner Distribution. Sov. Phys. Dokl. **1**, 414–418 (1956)
18. Wigner, E.: On the quantum correction for thermodynamic equilibrium. Phys. Rev. **40**, 749–759 (1932)
19. Zamanian, J., Stefan, M., Marklund, M. and Brodin, G.: From extended phase space dynamics to fluid theory. Phys. Plasmas **17**, 102109–102113 (2010)
20. Zhou, J. and Ferry, D. K.: 2-D Simulation of quantum effects in small semiconductor devices using quantum hydrodynamic equations. VLSI Design **3**, 159–177 (1995)
21. Zutic, I., Fabian J. and Das Sarma, S.: Spintronics: fundamentals and applications. Rev. Mod. Phys. **76**, 323–410 (2004)

Index

A
Adiabatic coefficient, 84
Adiabatic equation of state, 11, 194
Adiabatic local density approximation
    (ALDA), 32
Adiabatic speed of sound, 119
Alfvén velocity, 120, 128, 129

B
Bernstein–Greene–Kruskal (BGK)
    equilibrium, 44
Bessel function, 123
Bohm–Gross dispersion relation, 35, 194
Bohm potential, 42, 72, 73, 82, 98, 103,
    109, 114, 115, 119, 121, 126, 129,
    131, 134, 135, 137, 144, 148, 162,
    170, 171, 189, 192
Bright soliton, 101

C
Caldeira–Legget model, 33
Canonical partition function, 88
Classical ion-acoustic velocity, 97
Coulomb gauge, 110, 111, 199

D
Dark soliton, 101, 153
De Broglie wavelength, 3, 8, 9, 32
Debye length, 5, 8, 134
Debye sphere, 8
Decay instability, 139–142, 169
Degeneracy parameter, 8, 9
Dielectric function, 54, 55

E
Energy coupling parameter, 8–9, 30, 32
Enthalpy, 73, 122

F
Fermi–Dirac distribution, 6, 76, 98, 189
Fermi energy, 5, 7, 8, 31, 98, 123, 139, 142
Fermi gas, 35, 66, 72, 74, 81–85, 88, 89, 91,
    96, 123, 127, 135, 193
Fermi momentum, 7
Fermi sphere, 8
Fermi temperature, 7, 9, 96, 193
Fermi velocity, 5, 7, 74, 75, 78, 82, 84, 85, 87,
    96, 98, 123, 135

G
Gauge invariant Wigner function (GIWF),
    195–197, 199, 202
Gauge transformation, 112, 177, 196

H
Harris sheet solution, 125–130
Heat flux vector, 191

I
Ideal quantum magnetohydrodynamic model,
    119–121
Ion cyclotron velocity, 120
Isothermal equation of state, 127, 128

K
Kinetic pressure, 69, 70, 74, 125, 129, 189
Kinetic velocity, 68, 70, 71, 82
Korteweg–de Vries equation, 99–103

F. Haas, *Quantum Plasmas: An Hydrodynamic Approach*, Springer Series on Atomic,
Optical, and Plasma Physics 65, DOI 10.1007/978-1-4419-8201-8,
© Springer Science+Business Media, LLC 2011